Raphael Liesner

PhiC31 integrase as a promising tool in nonviral gene therapy

Raphael Liesner

PhiC31 integrase as a promising tool in nonviral gene therapy

Südwestdeutscher Verlag für Hochschulschriften

Imprint
Any brand names and product names mentioned in this book are subject to trademark, brand or patent protection and are trademarks or registered trademarks of their respective holders. The use of brand names, product names, common names, trade names, product descriptions etc. even without a particular marking in this work is in no way to be construed to mean that such names may be regarded as unrestricted in respect of trademark and brand protection legislation and could thus be used by anyone.

Publisher:
Südwestdeutscher Verlag für Hochschulschriften
is a trademark of
Dodo Books Indian Ocean Ltd., member of the OmniScriptum S.R.L Publishing group
str. A.Russo 15, of. 61, Chisinau-2068, Republic of Moldova Europe
Printed at: see last page
ISBN: 978-3-8381-2350-9

Zugl. / Approved by: München, Ludwig-Maximilians-Universität, Dissertation, 2010

Copyright © Raphael Liesner
Copyright © 2011 Dodo Books Indian Ocean Ltd., member of the OmniScriptum S.R.L Publishing group

Table of contents

Table of contents i

List of Tables iv

List of Figures iv

Abbreviations vi

Deutsche Zusammenfassung 1

1. Introduction 2
   1.1   Gene therapy 2
   1.2   Gene transfer systems 4
   1.3   Non-viral vectors 6
   1.3.1 Episomally persisting non-viral vectors 6
   1.3.2 DNA recombination-mediated integration and excision by transposases and recombinases 7
   1.4   PhiC31 integrase 12
   1.4.1 Recombination and integration efficiency 12
   1.4.2 Integration specificity and aberrant integration events 14
   1.4.3 Domain organisation and putative protein structure of the PhiC31 integrase 16
   1.4.4 Applications for PhiC31 integrase 20
   1.4.5 Attempts to improve PhiC31 integrase integration efficacy 20
   1.5   Aims of this study 22

2. Material 23
   2.1   Laboratory equipment 23
   2.2   Chemicals 23
   2.3   Enzymes 24
   2.3.1 Restriction endonucleases 24
   2.3.2 Other enzymes 24
   2.4   Solutions, media and buffers 25
   2.4.1 Media, supplements and reagents for eukaryotic cell lines 27
   2.5   Kits 28
   2.6   Organisms 29
   2.6.1 Bacteria 29
   2.6.2 Eukaryotic cells 29
   2.7   Oligonucleotides 30
   2.8   Plasmids 33

Table of Contents

3. Methods 35

3.1 General methods with eukaryotic cell culture 35
3.1.1 Cultivation of eukaryotic cells 35
3.1.2 Storage of eukaryotic cells 35
3.1.3 Cell counting of cultured eukaryotic cells 35
3.1.4 Transfection of eukaryotic cells with plasmid DNA 35
3.1.5 Transfection with siRNA and plasmid DNA and further applications 36
3.2 Process to select stable cell lines with integrated plasmid 38
3.2.1 Quantification of integration events 39
3.2.2 Establishment of stably transfected cell lines 40
3.2.3 Isolation of genomic DNA from eukaryotic cells 40
3.2.4 Analysis of integration events by plasmid rescue 41
3.2.5 Estimation of the quantity of colonies dependent on the integrase plasmid
concentration 42
3.3 Molecular biology techniques 43
3.3.1 Strain cultivation and storage of bacteria 43
3.3.2 Transformation of bacteria 43
3.3.3 Preparation of plasmid DNA 45
3.3.4 Polymerase chain reaction (PCR) 45
3.3.5 Design of a linker sequence and construction of a cloning vector pCS+NotI 48
3.3.6 Restriction digestion of pDNA and gel electrophoresis 48
3.3.7 Isolation of DNA fragments from agarose gels 49
3.3.8 Dephosphorylation of DNA fragments 49
3.3.9 Ligation of DNA fragments and vectors 49
3.3.10 Determination of DNA concentration 49
3.3.11 DNA sequencing and DNA alignments 50
3.3.12 Analysis of RNA, isolation and reverse transcription into cDNA 50
3.3.13 Quantitative real-time PCR to determine relative DAXX knock down 50
3.4 Fluorescence activated cell sorting (FACS) 51
3.5 Dual luciferase assay 52
3.6 Animal studies 53
3.6.1 Hydrodynamic tail vein injection 53
3.6.2 Measurement of alanine aminotransferase (ALT) 54
3.6.3 Enzyme Linked Immunoabsorbent Assay (ELISA) 54

4. Results 55

4.1 Construction of integrase mutants by site-directed mutagenesis 55
4.1.1 Preparation of truncated cloning vector pCS+NotI 56
4.1.2 Generation of mutated integrase binding domain by overlapping PCR 57
4.1.3 Generation of plasmids encoding the mutated integrase gene 58
4.2 Integration efficiency in HeLa cells of PhiC31 integrase mutant derivatives 59
4.3 Dose dependent studies with different amounts of PhiC31 integrase plasmids . 61
4.4 The effect of double mutants on integration efficiency 64
4.5 Integration efficiencies of integrase mutants in cell lines of different origin 66
4.5.1 Integration efficiencies in HCT cells 66
4.5.2 Integration efficiencies in Huh7 cells 67
4.5.3 Integration efficiencies in HEK 293 cells 68
4.5.4 Integration efficiencies in Hep1A cells 69
4.5.5 Evaluation of integration specificity in DAXX siRNA transfected cells 70

4.6	Integrase mediated excision in context of chromosomal DNA	73
4.6.1	Establishment and evaluation of GFP reporter cell lines	74
4.6.2	Evaluation of PhiC31 mutants in the eGFP reporter cell line #28	75
4.7	Integrase mediated excision within episomal plasmid DNA	78
4.7.1	An extrachromosomal assay to measure PhiC31 integrase mediated integration	78
4.7.2	Construction of reporter plasmids containing favoured pseudo attP sites	81
4.7.3	Evaluation of integrase mediated site-specific excision at attB/pseudo attP	83
4.8	Integration specificity of PhiC31 integrase in the genome of HCT cells	88
4.9	Human Factor IX (hFIX) expression upon plasmid delivery in murine liver	92

5. Summary 97

6. Discussion 98

6.1	Mutagenesis and other approaches to improve recombination of PhiC31 integrase	98
6.2	PhiC31 integration efficiency	101
6.3	Excision activity of PhiC31 integrase mutants	103
6.4	Specificity of PhiC31 integrase	104
6.5	Evaluation of integration efficiency in murine liver	107
6.6	Outlook and future perspectives	109

7. References 111

8. Appendix 125

Selection of amino acids discussed within this study. 125

Publications 126

Acknowledgements 127

List of Tables

Table 1.1	Number of gene therapy clinical trials...	2
Table 1.2	Representative SSR family members grouped in two distinct families based on tyrosine and serine mediated catalysis.............................	10
Table 2.1	Bacterial strains...	29
Table 2.2	Eukaryotic cell lines...	29
Table 2.3	Oligonucleotides used for site-directed mutagenesis of the PhiC31 integrase...	30/31
Table 2.4	Oligonucleotides used for cloning and sequencing........................	32
Table 2.5	Duplex RNAused for quantitative real-time PCR.............................	32
Table 2.6	Plasmids used for cloning and transfection...................................	33/34
Table 3.1	Guidelines for transfection reagents in different well size using FuGENE6 transfection reagent:DNA at a ratio of 3:1......................	36
Table 3.2	Transfection setup and applications...	37
Table 3.3	Overview over transfection setups with increasing and decreasing integrase plasmid amounts...	43
Table 3.4	Temperature profiles of the first PCR...	47
Table 3.5	Temperature profiles for overlapping PCR using proofreading Pfx and KOD polymerases...	47
Table 3.6	Experimental outline of transfection conditions of luciferase assays performed..	53
Table 4.1	Constructed PhiC31 integrase mutants..	60
Table 4.2	Overview of PhiC31 integrase-mediated recombination activities in colony-forming assay, FACS and luciferase assay carried out *in vitro*..	87
Table 4.3	Summary of rescued sites of PhiC31 integration in HCT cell line.........	91
Table 4.4	Experimental setup of two independent *in vivo* experiments...............	93
Table 4.5	Serum hFIX concentrations from different groups of mice injected with PhiC31 integrase plasmid and hFIXmg plasmid post initial CCl_4 injections...	96
Table 8.1	Selection of amino acids discussed within this study........................	126

List of Figures

Figure 1.1	The putative mechanism of recombination by a serine recombinase....	11
Figure 1.2	Integration mediated by PhiC31 integrase.......................................	13
Figure 1.3	PhiC31 integrase-specific attachment sequences............................	13
Figure 1.4	Domain organisation of gamma delta resolvase and PhiC31 integrase	17
Figure 1.5	Secondary structure prediction of the PhiC31 integrase protein..........	18
Figure 1.6	Protein structures of the gamma delta resolvase..............................	19
Figure 3.1	Overview of process after with siRNA and plasmid DNA...................	37
Figure 3.2	The plasmids used for co-transfection into cell lines.........................	38
Figure 3.3	Outline of plasmid transfection, cultivation and subsequent applications	39
Figure 3.4	The plasmid rescue..	42
Figure 3.5	Overview of two-step overlapping PCR to generate point mutations....	48
Figure 4.1	Cloning strategy for the construction of integrase point mutants.........	56

List of Figures

Figure 4.2 Agarose gel electrophoresis of the vector backbone pCS+*NotI* uncut, B*am*HI digested, and *NotI*+*PstI* digested and respective vector maps. 57

Figure 4.3 PCR products obtained after the first round of PCR with mutagenic primers analysed by agarose gel electrophoresis............................ 58

Figure 4.4 Analysis of second round PCR products by agarose gel electrophoresis... 58

Figure 4.5 Nucleotide sequence and amino acid sequence of the integrase binding domain.. 59

Figure 4.6 Schematic overview over a colony-forming assay (CFA).................... 61

Figure 4.7 Integration efficiency of all constructed integrase mutants obtained by colony-forming assay (CFA) in HeLa cells.................................. 62

Figure 4.8 Dose dependent studies with increasing plasmid transfection ratios of selected integrase mutants... 64

Figure 4.9 Dose dependent studies with decreasing amounts of integrase plasmid compared to substrate plasmid p7....................................... 65

Figure 4.10 Integration efficiencies of single and double mutants in HeLa cells...... 66

Figure 4.11 Integration efficiencies of double mutants in HeLa cells at two different plasmid ratios p7:Int = 1:0.5 and 1:20............................. 67

Figure 4.12 Integration efficiencies of selected integrase mutants in HCT cells...... 69

Figure 4.13 Integration efficiencies of selected integrase mutants in Huh7 cells...... 69

Figure 4.14 Integration efficiencies of selected integrase mutants in 293 cells....... 70

Figure 4.15 Integration efficiencies of selected integrase mutants in Hep1A cell line.. 71

Figure 4.16 The relative knock down of DAXX by DAXX-specific siRNA in relation to nonspecific siRNA.. 73

Figure 4.17 Integration efficiency of integrase in siRNA transfected 293 cells........ 74

Figure 4.18 The p-*attP*-polyA-*attB*-eGFP reporter plasmid.................................. 75

Figure 4.19 FACS analysis of various clones upon transfection of integrase plasmids... 77

Figure 4.20 FACS analysis in reporter cell line #28... 79

Figure 4.21 Integrase mediated excision activity of integrase mutants.................. 80

Figure 4.22 Schematic overview over luciferase expression after PhiC31 integrase-mediated polyA excision.. 81

Figure 4.23 Excision activity of integrase mutants detected by a luciferase assay in 293 cells... 82

Figure 4.24 Excision activity of integrase mutants detected in a luciferase assay... 83

Figure 4.25 Analysis of amplified genomic hot spot DNA sequences with preferred PhiC31 integrase sites by agarose gel electrophoresis..................... 84

Figure 4.26 Plasmid constructs with wt *attP* site and three different *pseudo attP* sites... 85

Figure 4.27 Excision activity of integrase mutants at *pseudo attP* site 2q.11.......... 86

Figure 4.28 Excision activity of integrase mutants at *pseudo attP* site 12q.22........ 87

Figure 4.29 Excision activity of integrase mutants at *pseudo attP* site 19q13.31..... 88

Figure 4.30 Positions of clonally rescued PhiC31 integration sites in the human genome.. 92

Figure 4.31 DNA sequences injected into female C57BL/6 mice............................ 95

Figure 4.32 Surveillance of alanine transaminase (ALT) levels to evaluate liver damage.. 96

Figure 4.33 Human Factor IX (hFIX) expression levels *in vivo*............................ 97

Figure 8.1 Selection of amino acids discussed within this study........................ 126

List of Abbreviations

A	Alanine
AA	Amino acid
AAV	Adeno-associated virus
ADA	Alanine deaminase
Adv	Adeno virus
ALT	Alanine transaminase
Amp	Ampicillin
att site	Attachment site
BD	PhiC31 integrase binding domain
bp	Base pairs
CCl_4	Carbon tetrachloride
cDNA	Complementary DNA
CFA	Colony-forming assay
Chr	Chromosome
CIP	Calf Intestinal Alkaline Phosphatase
cm	Centimetre
CMV-p	Cytomegalovirus promoter
Cp	Crossing point
Cre	cyclisation recombination
CTD	C-terminal domain
D	Aspartic acid
dATP	Deoxy adenosine triphospahte
DMEM	Dulbecco´s modified Eagle medium
DMSO	Dimethyl sulfoxide
DNA	Deoxyribonucleic acid
dNTPs	Deoxyribonucleotide triphosphates (dATP, dCTP, dTTP, dGTP)
D-PBS	Dulbecco´s Phosphate buffered saline
DSMZ*	Deutsche Sammlung von Mikroorganismen und Zellkulturen (German Collection of Microorganisms and Cell Cultures)
E	glutamic acid
EBV	Epstein Barr virus

List of Abbreviations

EDTA	Ethylenediaminetetraacetic acid
eGFP	Enhanced green fluorescent protein
ELISA	Enzyme-linked immunosorbent assay
EtBr	Ethidium bromide
FACS	Fluorescence activated cell sorting
FBS	Foetal bovine serum
FLP	Flippase
FRT	Flippase recognition target
γδ resolvase	Gamma delta resolvase
GFP	Green fluorescent protein
hFIX (mg)	Human blood coagulation factor IX (mini gene)
HRPO	Horseradish peroxidase
hs	Hot spot
HSV	Herpes simplex virus
Int	PhiC31 integrase
IR	Inverted repeates
K	Lysine
Kan	Kanamycin
Kb(p)	Kilo base (pairs)
kDa	Kilo Dalton
l	Litre
LB medium	Luria Bertani medium
Luc	Luciferase
loxP	L̲ocus of X̲-over in P̲1
MLV	Murine leukemia virus
ml	Millilitre
min	Minutes
mInt	Mutant integrase
mRNA	Messenger RNA
NEB	New England Biolabs
neo[R]	Neomycin resistance gene
NLS	Nuclear localisation signal
nm	Nanometre
NTD	N-terminal domain

List of Abbreviations

OD_{260}	Optical density at 260 nm
ORF	Open reading frame
oriP	Origin of replication
PBS	Phosphate buffered saline
PCI	Phenyl-chloroform-isoamylalcohol
PCR	Polymerase chain reaction
pDNA	Plasmid DNA
PNK	Phosphate nucleotide kinase
R	Arginine
RLU	Relative light units
RNA	Ribonucleic acid
rpm	Rotations per minute
SB transposase	Sleeping Beauty transposase
SDS	Sodium dodecyl sulphate
SV 40-p	Simian virus 40 promoter
RT PCR	Real-time polymerase chain reaction
SCID	Severe combined immune deficiency
SOC	Super Optimal Broth
SSR(s)	Site-specific recombinase(s)
TAE	Tris-acetate-EDTA
TBS-T	Tris buffered saline-Tween 20
TE	Tris EDTA
TRIS	Tris-(hydroxymethyl)-ammonium methane
wt	Wild type
µl	Microlitre

Deutsche Zusammenfassung

Die Bakteriophagen Integrase PhiC31 stellt ein viel versprechendes Werkzeug zur Integration genetischen Materials im nicht viral-basierten Gentransfer dar. Die PhiC31 Integrase vermittelt die Rekombination von spezifische Erkennungssequenz *attB* enthaltenden Plasmiden mit natürlich vorkommenden *attP* Erkennungssequenzen innerhalb des Zielgenoms mit unterschiedlicher Integrationsfrequenz und -spezifität. Nebeneffekte der Integration in Form von insertioneller Mutagenese, z.B. große Deletionen und chromosomale Veränderungen im Genom der Zielzelle konnten beobachtet werden.

Ziel dieser Dissertation war, die PhiC31 Integrase vermittelte Effizienz zu verbessern und die Spezifität zu adressieren. Als Ansatz wurde die Mutagenese der DNA-Bindungsdomäne der Integrase basierend auf Punktmutanten zur verbesserten Integrationseffizienz gewählt. Integrationsassays wurden in verschiedenen humanen Zelllinien durchgeführt. Etablierung von Doppelmutanten, sowie Dosisoptimierung des Integrase kodierenden Plasmids verbesserten die Integrationseffizienz mehr als dreifach, verglichen mit der Wildtypintegrase in den Zelllinien HeLa und HCT. Weitere Assays verglichen die Exzisionsaktivität der Integrasemutanten mit dem Wildtyp. Bei fünf Mutanten wurde eine etwa zweifach erhöhte Exzision gefunden. Die Beurteilung der Spezifität der Integrasemutanten erfolgte durch Substitution der Wildtyp-*attP* Sequenz mit drei *Pseudo attP* Erkennungssequenzen des Reporterplasmids, deren erhöhte Spezifität bereits dokumentiert war. Einzelne Mutanten zeigten eine zweifach erhöhte Exzisionsaktivität. Die Rekombinationsaktivität von Integrasemutanten wurde im Kontext chromosomaler DNA mittels einer stabil GFP-exprimierenden Reporter-zelllinie, in der die eGFP Expression mittels Integrase-vermittelter „Raus-Rekombination" eines polyA Stoppsignals angeschaltet wird, untersucht. Auf chromosomaler Ebene wurde keine verbesserte Ausschneidungsaktivität erreicht. Zur Evaluierung der *in vivo* Effizienz zweier ausgewählter PhiC31 Integrasemutanten, die *in vitro* erhöhte Integrationsaktivität aufwiesen, wurden zwei Plasmide am C57BL/6 Mausmodell getestet. Reportergen war ein für den humanen Koagulations-Faktor IX kodierendes Gen. In Abhängigkeit von der Integrationseffizienz der Mutanten und des Wildtyps, wurden im Zeitraum von 100 Tagen ähnliche Expressionslevel gefunden. Die Mutanten zeigten keine Verbesserung der Langzeitexpression von humanem Faktor IX. Die hier durchgeführten Studien zur Mutationsanalyse der Phagenintegrase PhiC31 zeigten einen wirksamen Ansatz zur Verbesserung der PhiC31 Integrase-vermittelten Integrationseffizienz *in vitro*.

1. Introduction

1.1 Gene therapy

Historically, gene therapy can be defined as the treatment of hereditary and metabolic diseases or cancer by introducing therapeutic genes into the target cells of affected tissues (Friedmann and Roblin, 1972). The function of the inserted gene is thereby able to compensate for a genetic malfunction or deficiency of the mutated gene. Gene therapists distinguish between somatic gene therapy and germline gene therapy. Somatic gene therapy is considered to be the traditional treatment in gene therapy. This form of therapy addresses treatment of inherited and acquired genetic diseases within one individual by selective transfer of normal genes into certain somatic target cells. Germline based gene therapy deals with the genetic alteration of germ cells such as ovules and sperm cells and their progenitor cells. The latter therapy enables introducing genetic changes into the germ line, which are then passed on to future generations (Nielsen, 1997).

a) Gene therapy trials

The concept of somatic gene therapy has become accepted to date especially in animal models. Gene therapy vectors are applied to the patients target tissue by different methods. The *in vivo* based gene therapy refers to the transfer of a foreign gene into the targeted, living tissue of a whole organism. *Ex vivo* gene therapy involves the removal of living cells or tissues from an organism, subsequent manipulation such as foreign gene insertion *in vitro* and re-implantation into the organism. This form of gene therapy is used for treating inherited diseases or in bone marrow transplants. Gene therapy clinical trials have been approved and documented since 1989. During the last twenty years 1644 clinical trials had been documented. The four mostly diagnosed indications are presented in Table 1.1.

Table 1.1. Number of gene therapy clinical trials. [1]

Indication	Numbers of Trials
Cancer	1060 (64.5 %)
Cardiovascular disease	143 (8.7 %)
Monogenic disease	134 (8.2 %)
Infectious disease	131 (8.0 %)

[1] http://www.wiley.com/legacy/wileychi/genmed/clinical/ provided by the Journal of Gene Medicine, (06/2010)

Introduction

Despite current difficulties, gene therapy has great potential to address, treat and cure infections and inherited diseases, based on monogenic malfunctions. The cure for cancer, in which several genes are affected (multi-genetic) by means of gene therapy, is still a big challenge. However, many questions remain open concerning risks and benefits of gene therapeutic approaches, which will be objectives for future investigations.

b) Clinical trials: progress and problems

Since the start of gene therapy trials the number of human genes associated with disease states and vector systems have been steadily increasing. Outstanding successes in human gene therapy in curing diseases such as SCID (severe combined immune deficiency) by retroviral vector-mediated gene transfer have been achieved. The treatment of X-linked SCID, an inherited disorder caused by γC cytokine receptor deficiency was first reported by Cavazzana-Calvo et al. (2000). For the first time, alanine deaminase (ADA)-SCID, which provokes a defect in the purine metabolism by defective T and B cell function was successfully treated in ADA SCID patients (Aiuti et al., 2002). Currently, eight out of ten patients do not require enzyme-replacement therapy anymore since the patient's blood cells express ADA by themselves (Aiuti et al., 2009).

However, failures and severe side-effects during phase I clinical trials have raised concerns about safety issues due to immune inflammatory response and insertional mutagenesis leading to prompt death or leukaemia more than two years after the administration of adenoviral or retroviral vectors, respectively (Raper et al., 2003; Check, 2002; Hacein-Bey-Abina et al., 2003a). The success of human gene therapy trials has been seriously queried upon a tragic setback. During a phase I clinical trial to test the safety of using a second-generation ΔE1/E4 adenovirus vector, a young man died (Marshall, 1999; Bostanci, 2002; Thomas et al., 2003). The patient suffered from a partial deficiency of ornithine transcarbamylase (OTC), a liver enzyme needed for removal of excessive nitrogen from amino acids and proteins. This patient received the highest dose of vector (3.8×10^{13} particles) in the study via transduction of an adenovirus carrying the gene encoding for OTC. Within four days of treatment the patient died from multiorgan failure (Bostanci, 2002).

The first *ex vivo* gene therapy study was performed to treat X-SCID in infants in Paris in 1999 by means of retroviral vector transduction (Cavazzana-Calvo et al., 2000). A similar gene therapy phase I trial had started in the United Kingdom (Gaspar et al., 2004). Haematopoietic stem cells were transduced *ex vivo* with replication deficient recombinant murine leukaemia virus (MLV) vectors carrying the therapeutic gene encoding common

Introduction

cytokine-receptor chain (γc) to restore the missing IL-2 receptor γ (IL2RG). Yet, the clinical benefit of gene therapy trials was hampered by the development of leukaemia in four out of nine young patients more than 2.5 years after initial retrovirus-based gene delivery (Hacein-Bey-Abina et al., 2003b; Hacein-Bey-Abina et al., 2008). Retroviral vector integration near or into the proviral integration site, LIM domain only 2 (LMO2), or into different proto-oncogenes activated transcription of cancer related genes and lead to chromosomal translocations (Hacein-Bey-Abina et al., 2008). Recently, a 3-year-old boy treated in a hospital in London, also developed a chronic bone marrow disease two years after retrovirus vector mediated treatment for X-SCID (Cole, 2008).

The latest moderately successful gene therapeutic treatment was reported while medicating patients suffering from an inherited blinding disease called Leber`s congenital amaurosis (LCA) (Bainbridge et al, 2008; Maguire et al., 2008). The patients' blindness has onset during childhood caused by irreversible retina degeneration through a mutation in the retinal pigment epithelium (RPE65)-specific 65-kilo Dalton (kDa) protein gene, which is implicated in the production of vitamin A as a rhodopsin precursor. Rhodopsin is required for vision and photoreceptor function. In two trials each with three patients, the patients were subretinally injected with a recombinant adeno-associated virus (AAV) vector encoding the RPE65 gene. Modest improvement in visual function and acuity could be achieved in four out of six patients without adverse events. However, additional studies need to be carried out to support the potential for visual restoration using the current practise.

1.2 Gene transfer systems

Several different vectors and methods have been used, which are capable of transferring therapeutic nucleic acids into the cells. In principal two major groups exist: viral vectors and non-viral systems. The latter are also referred to as synthetic vector systems, which consist generally of a carrier, a compound that mimics the function of viral capsids, and one or more plasmids encoding the therapeutic gene and possibly different genes encoding integration-supporting enzymes.

Two-thirds of all gene therapy vectors used in clinical trials are viral vectors (Edelstein et al., 2007). The vectors are mainly composed of retroviruses and adenoviruses (AdVs) which are able to mediate highly efficient gene transfer (Ragot et al., 1993; Annenkov et al., 2002). The AAV is the smallest among the viruses used for gene transfer (Dong et al., 1996) and is non-inflammatory and non-pathogenic (Nakai et al., 1999).

Introduction

Viral vectors differ in their vector genome forms. AdV and HSV genomes persist mainly in the nucleus, AAV genomes remain preferentially episomal (Nakai et al., 2001), although AAV integration in *in vivo* studies has also been documented as well (Nakai et al., 2003a; Miao et al., 1998). The group of retroviruses integrate their genomes into the host chromatin (Mitchell et al., 2004; Schroeder et al., 2002).

For therapeutic applications, inactive replication-deficient viruses or viral capsids are used. Parts of the viral genome are deleted and replaced by an expression cassette including therapeutic genes of choice (Alba et al., 2005; Jaeger and Ehrhardt, 2007). The tropism of a few viral vectors as Lentiviruses, AAVs and AdVs is generally multi-faced. But inflammatory herpes simplex virus (HSV)-1 exhibits a strong preference for neurons (Palmer et al., 2000), whereas retroviruses transduce only dividing cells.

While viral vectors are able to mediate highly efficient gene transfer, there are several drawbacks to using them. For example, most viral vectors mediate an inflammatory response triggering immunogenic complications, in particular AdV (Thomas et al., 2001; Simon et al., 1993; Kafri et al., 1998) and HSV-1 (Epstein et al., 2005). Compared to synthetic vector systems, all viral vectors have a limiting cargo capacity. Additionally, the preparation of recombinant viral vectors in terms of production and purification is laborious and complex, especially for AdV-based vectors (Palmer and Ng, 2003). Moreover, viral vectors are frequently accompanied by various technical difficulties, e.g. availability for optimal therapeutic doses, biomedical safety or immunological properties of the individual patient (Lundstrom and Boulikas, 2003; Levine, 1987).

In general, non-viral gene transfer systems represent an appropriate alternative to viral transduction with lower toxicity and lower immunogenicity than viral systems. These carriers, either synthetic or naturally based, are able to shield their cargo in form of plasmids or si (small interfering) RNA and are used to deliver therapeutic DNA into cells. The methods of non-viral gene delivery can be subdivided into physical (carrier-free) delivery and chemical approaches using synthetic vectors. The physical approaches involve needle injection, electroporation, gene gun and hydrodynamic delivery (Wolff et al., 1990; Heller et al., 2005; Yang and Sun, 1995). Transfection of hepatocytes, mediated by hydrodynamic intravascular injection of naked DNA via the tail vein of mice has become routine method for nucleic acid delivery into the liver (Gao et al., 2007; Zhang et al., 1999). The chemical approaches include polymer-based transfection or lipofection (Liu et al., 2003). The carriers consist of synthetic or native compounds as cationic lipids or polymers for transgene delivery (Gao et al., 2007). A successful gene transfer is strongly dependent on efficient delivery into the target cells although several hurdles facing the gene delivery

Introduction

via extracellular and intracellular environments (Pouton and Seymour, 2001) thereby inhibiting the route of the genetic cargo.

1.3 Non-viral vectors

There are several non-viral vector systems, e.g. viral Epstein-Barr virus (EBV)-based plasmid replicons and pEPI-based vectors. Several of these systems have been improved in the last decade. Non-viral vectors have shown increasing gene therapeutic potential and are of upcoming importance as a strong alternative to viral vectors. Low cost, simplicity of use, ease of large-scale production and lack of specific immune response are beneficial properties of non-viral vectors (Niidome and Huang, 2002). Like viral vectors, there are drawbacks with non-viral vectors as well. One of the major drawbacks is the low *in vivo* and *in vitro* transfection efficiency due to versatile cellular barriers, which prevent proper delivery of DNA into the target cells. In order to overcome this, viral properties are transferred into non-viral systems in order to integrate or replicate the vectors and to achieve sustained and stabilised long-term transgene expression (Glover et al., 2005).

Non-viral vectors for gene transfer are divided into non-integrating or episomally persisting elements and integrating vectors. Episomally persisting vectors comprise minicircles including a scaffold matrix attachment region (S/MAR) and pEPIs (Piechaczek et al., 1999; Chen et al., 2005; Darquet et al., 1997; Nehlsen et al., 2006). These plasmid-based vectors are still being optimised in respect to safety and efficient autonomous replication so as to improved stable long-term extrachromosomal persistence (Mairhofer and Grabherr, 2008). Non-viral vectors capable of somatic integration of entire plasmids or linear DNA sequences are plasmids that carry genes encoding various transposon systems derived from different eukaryotes and several bacteriophage-derived site-specific recombinases (SSR) such as integrases.

1.3.1 Episomally persisting non-viral vectors

Episomal vector systems with non-viral origins contain viral plasmid replicons, e.g. the origin of replication (oriP), Epstein-Barr virus nuclear antigen 1 (EBNA1), or chromosomal elements (Conese et al., 2004). In contrast to integrating vectors, episomal non-viral vectors do not cause insertional mutagenesis as a severe side effect of integration. Transfected plasmids only persist as extrachromosomal entities in rapidly dividing cells if they are able to replicate or if they are able to utilise the cellular mechanism for their nuclear retention. Another strategy to stabilise the plasmid's presence in the nucleus is to modify the bacterial backbone to render the plasmid replication-competent. EBV-derived episomal persisting, self-replicating vectors, in which viral and non-viral elements are combined into one vector plasmid, overcome the rate-limiting step of non-viral gene transfer (Yates and Guan, 1991). Increased levels of transgene expression and improved transfection efficiency might be obtainable with EBV-based episomal vectors (Mazda et al., 1997). The first small episomal vectors, which did not require any viral encoded trans-acting factors for replication were the pEPI vectors (Piechaczek et al., 1999). pEPI-1 replicates episomally at a copy number of about ten in Chinese hamster ovary (CHO) cells and is stable in the absence of selection for over one hundred generations ensuring long-term gene expression (Piechaczek et al., 1999).

1.3.2 DNA recombination-mediated integration and excision by transposases and recombinases

Being a safer alternative to recombinant integrating viral vectors, non-viral gene transfer systems have gained importance. The two major non-viral gene transfer vectors used for somatic integration are based on the Sleeping Beauty (SB) transposase and the bacteriophage PhiC31 integrase system.

The **SB transposase** is a transposable element and belongs to the Tc1/mariner superfamily of transposons derived from fish (Ivics et al., 1997). Transposons are mobile DNA sequences that are able to move within a genome. The SB transposase system is utilised for insertion of an *in trans* delivered transposon with a cargo capacity of about 10 kb (Zayed et al., 2004), in which the transgene is flanked by 250-bp long terminal inverted repeats (IR). Both the transposase and the inverted repeats constitute the SB transposon system (Ivics et al., 1997). At the IRs, which represent the target sites, the precise cut-and-paste mechanism takes place resulting in transposon integration into the chromosome (Ivics et al., 1997; Izsvák et al., 2002). In the SB transposon system, the quantity of transposase is limited by an effect termed overproduction inhibition. Transposition efficacy is negatively affected when the transposase dose reaches a certain threshold (Mikkelsen et al., 2003; Yant et al., 2000). A cellular mechanism described as postintegrative gene silencing has been documented in human cells. Transposon silencing was proposed to be associated with DNA methylation, histone deacetylation and promoter and cell line dependent (Garrison et al., 2007). Although the consensus sequence of SB target sites was found to be a short thymidine adenosine (TA) dinucleotide, the selection of the target insertion site is primarily determined by the level of DNA structure (Vigdal et al., 2002). Transcription units and their upstream regulatory sequences are strongly preferred by SB

Introduction

transposases (Yant et al., 2005). Several mutational screens within the nucleotide sequence of the SB transposase increased the efficiency of SB between twofold and 100-fold in the last five years (Geurts et al., 2003; Yant et al., 2004; Zayed et al., 2004; Mátés et al., 2009).

Since the sequence to be transposed consists of therapeutic genes, the SB transposase system has been frequently used as a versatile tool for efficient gene delivery in several preclinical gene therapy applications (Liu et al., 2006a) to achieve long-term therapeutic transgene expression in lung and liver cells of mice (Yant et al., 2000; Belur et al., 2003).

The first clinical trial including a SB transposon vector has been approved, in which genetically altered T cells will be transferred into patients with $CD19^+$ B lymphoid malignancies in an *ex vivo* approach (Williams, 2008). Besides applications in gene therapy or gene transfer, the SB transposase system is used extensively for cancer gene discovery in combination with insertional mutagenesis, as well as germline transgenesis, and functional genomics (Collier et al., 2005; Izsvák and Ivics, 2004; Luo et al., 1998).

The **PhiC31 integrase** belongs to the serine-catalysed invertase/resolvase family of recombinases (Stark et al., 1992; reviewed by Groth and Calos, 2004). SSRs recognise sequence specific target sites in length of 30-40 base pairs (bp) and mediate recombination between these so-called attachment sites, *attP* and *attB*. This group of enzymes form two functionally and structurally diverse families (Smith and Thorpe, 2002). These families are based on their sequence homology, biochemical properties (Stark et al., 1992) and their mechanism of catalysis (Hatfull and Grindley, 1988). The main representatives of tyrosine- and serine-based integrase, all being recombination-efficient in mammalian cells, are listed in Table 1.2.

The **tyrosine recombinase or lambda integrase family** makes use of a conserved tyrosine catalytic residue to mediate covalent bonds between the recombinase and the DNA target sequence (Sorrell and Kolb, 2005). Well known representatives of the tyrosine-mediated recombinases are **c**yclisation **re**combination *(Cre)* recombinase from phage P1 and Flippase *(FLP)* recombinase from *Saccharomyces cerevisiae*. *Cre* mediates a bidirectional recombination between identical **lo**cus of **X**-over in **P**1 (loxP) sites (Hoess et al., 1990). After its discovery, the *Cre* loxP system has been widely used to manipulate eukaryotic and prokaryotic genomes (Sternberg, 1979). The *FLP* recombinase mediates inversion between two 34-bp **_Flp_** **r**ecognition **t**arget *(FRT)* sites arranged in opposite directions. Since the target site sequences of the SSR before and after the recombination reaction are identical, *Cre* and *Flp* mediated reactions are reversible, leading to subsequent excisions in the presence of recombinase. This renders the inserted DNA highly unstable. Contrary to the PhiC31 integrase specific attachment (*att*) sites, the recognition sites of both the *Cre* and *Flp* recombination systems are missing in the mammalian genome. Therefore, to create an integrating recombination-based system in mammalian cells previous stable insertion of the specific loxP or *FRT* recognition sites into the genome is required.

The second family of SSRs comprises the large, diverse and evolutionary unrelated invertase/resolvase or **serine-based recombinase family.** This family makes use of a conserved N-terminal serine residue to establish the covalent link between recombinase and DNA target sequence. Enzymes of this recombinase family utilise the hydroxyl group of the catalytic serine resulting in a strand exchange (Groth and Calos, 2004). The unidirectional site-specific recombination at the crossover point between two different DNA recognition sequences, the phage attachment site *attP* and the bacterial attachment site *attB* (Groth et al., 2000), results in the generation of two non-identical hybrid att sites, *attL* and *attR* (Rausch and Lehmann, 1991). Safe integration and long-term expression of the therapeutic gene is in theory an excellent premise for curing inherited diseases since life-long treatment is usually required. The PhiC31 integrase as the most studied representative among the serine catalysed recombinases represents a convenient novel tool for gene therapeutic interventions, and may provide desired properties and approaches to fulfil these requirements. From at least thirty members of the serine-based recombinase family, several serine based integrases have been studied for the potential of chromosomal engineering (Chen and Woo, 2008).

Among the serine recombinase family, the small gamma delta (γδ) resolvase, encoded by the transposon γδ, was the first to be solved in its crystal structure as a synaptic tetramer (Sauer, 1994). Since that time, several synaptic models of γδ resolvase were analysed in great detail (Rice and Steitz, 1994; Yang and Steitz, 1995; Sarkis et al., 2001; Li et al., 2005). Therefore, it can be described as the model recombinase for structural analysis, catalysis and biochemical properties. Related enzymes such as R4 integrase (Olivares et al., 2001), TP901-1 (Stoll et al., 2002), PhiFC1 (Yang et al., 2002) and PhiBT1 (Chen and Woo, 2005) were also investigated for recombination activity, yet with minor integration efficiency.

Introduction

Table 1.2. Representative SSR family members grouped in two distinct families based on tyrosine and serine mediated catalysis are listed.

	Phage name	Host	Amino acid length	Overlap region	Original reference
Tyrosine-based integrases	Lambda (λ)	E.coli	356	TTTATAC	Enquist et al., 1979
	HK022	E.coli	357	AGGTGAA	Yagil et al., 1989
	Cre (P1)	E.coli	343	ATGTATGC	Abremski and Hoess, 1984
	FLP	S. cerevisiae	423	TCTAGAAA	Andrews et al., 1985
Serine-based integrases	**PhiC31 integrase**	**Streptomyces lividans**	613 / 605	TTG	Kuhstoss and Rao, 1991
	R4	Streptomyces parvulus	469	GAAGCAGTGGTA	Groth et al., 2000
	TP901-1	Lactococcus lactis	485	TCAAT	Christiansen et al., 1996
	γδ	E.coli	183	TATTTATAAAT	Reed et al., 1982

The enzymatic understanding of the site-specific recombination mechanism of recombinase family members has been investigated for about two decades (Stark et al., 1989; Stark et al., 1992; Grindley, 1997; Li et al., 2005; reviewed by Grindley et al., 2006). The recombinase-mediated strand exchange requires supercoiled structures of both chromosomal and of plasmid DNA substrates. It also requires a recombinase dimer for a DNA strand or a tetramer for two strands while being recombinase-type dependent. This process involves binding of the resolvase dimer to the target DNA, their synapsis, double strand cleavage at the crossover points and subsequent exchange and rejoining of the DNA strands (Murley and Grindley, 1998). During catalysis, a synaptic complex of two DNA crossover sites and four recombinase subunits is formed (Grindley et al., 2006). Designed mutants, such as the serine recombinase Tn3, requires only short recombination sites and is a result of improved molecular understanding of synaptic and strand exchange processes (Burke et al., 2004; Arnold et al., 1999). Two different mechanistic models, the domain-swapping model and the subunit rotation model, for serine recombinase mediated strand exchange were proposed in which resolvase dimer structures composed of catalytic domains flexibly connected to DNA-binding domains exist (Rice, 2005). The domain-swapping model has not become accepted (Yang and Steitz, 1995; Craig et al., 2002). The majority of analysed serine-recombinase mediated recombination events supported the subunit rotation model (Figure 1.1) as the mechanism of choice for strand exchange (Li et al., 2005) for TP901-1, PhiC31 integrase and other related enzymes (Yuan et al., 2008).

Introduction

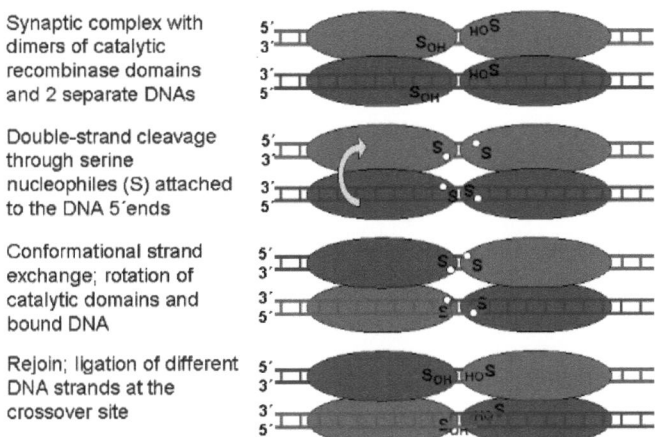

Figure 1.1. The putative mechanism of recombination by a serine recombinase.
The subunit rotation model for recombinase-mediated strand exchange is illustrated above. For simplicity only with the catalytic domains of the recombinase dimers (oval circles) synapsed to the DNA strand is shown. Both recombinase dimers and DNA strands are represented as upper and lower subunits, respectively. First, the synaptosome is formed by attaching the target DNA cleavage site to the catalytic subdomains (either as monomers or dimers). The hydroxyl group of the conserved catalytic serine residue nucleophilically attacks the DNA backbone, leaving free 3´OH groups (Stark et al., 1992). The strand exchange is accomplished by a 180° rotation of two half sites of the tetramer relative to each other. DNA strands subsequently rejoin and opposite strands are ligated. The figure is modified and derived from Grindley et al., 2006.

SSRs represent not only integration tools for plasmid-derived therapeutic genes delivered *in trans* but also are utilised for site-specific gene insertion into the genome of the target tissue. A second extensively used application of SSRs is based on intramolecular excision activity of a DNA sequence, which is flanked by both recognition sites. SSR mediated excision activity is utilised in applications such as gene activation, gene inactivation, and transgenesis e.g. with recombinase-mediated cassette exchange (RMCE). SSRs as an excision tool have become increasing relevant in the field of reverse genetics (Bischof and Basler, 2008).

In transgenesis the *FLP/FRT* system has been explored two decades ago in *Drosophila* genetics to perform *in vivo* manipulations (Golic and Lindquist, 1989). The technique of gene activation has been utilised to generate constitutive expression of transgenes in cells (Struhl and Basler, 1993). Transgene expression is turned on by the excision of an "*FLP* out" cassette. This sequence containing a marker gene flanked by recognition sequences (*FRT* sites) in the same orientation separates the transgene from its promoter and contains a transcriptional termination site (polyA signal) (Bischof and Basler, 2008). The removal consequently leads to the expression of the transgene. Gene inactivation by

Introduction

mitotic recombination between *FRT* sites involves the loss of a marker gene within a particular clone or a generated daughter cell and has been applied in *Drosophila* (Chou and Perrimon, 1992; Xu and Rubin, 1993). Transgenesis using the RCME strategy in combination with PhiC31 integrase has also been carried out in the fruit fly (Groth et al., 2004; Bateman et al., 2006). The genetic manipulations during cassette exchange were proved by the loss of the marker gene. Transgenesis with PhiC31 integrase has been optimised by the generation of cell lines with precisely mapped *attP* sites and by raising transformation efficiency (Bischof et al., 2007).

1.4 PhiC31 integrase

1.4.1 Recombination and integration efficiency

The temperate sensitive bacteriophage PhiC31 has a genome size of about 41.5 kbp (Harris et al., 1983) and infects a large number of streptomycetes, including *Streptomyces lividans* (Kuhstoss and Rao, 1991; Chater, 1986). The natural recombination process consists of a stable unidirectional and site-specific integration of the phage genome into the bacterial host genome (Figure 1.2.A). This property makes PhiC31 integrase an ideal tool for gene therapeutic applications (Calos, 2006). PhiC31 integrase was first shown to mediate efficient integration in a mammalian cell environment using extrachromosomal vectors in bimolecular integration assays (Groth et al., 2000). The mechanism of recombination (Figure 1.2.B) is carried out by covalent attachment of the transient DNA to the two integrase monomers. As observed in related integrase families, the serine-based integrase utilises the hydroxyl group of the catalytic serine residue to nucleophilically attack the DNA backbone (Stark et al., 1992). Serine integrases make a 2-bp staggered cut on both attachment sites, consequently breaking all four nucleotide strands at once. Double-stranded breaks made by protein dimers bound in *cis* follow a 180° rotation and subsequent DNA ligation (Groth and Calos, 2004). All four attachment sites involved in the recombination process, native *attP* and *attB* sites and the hybrid sites, *attL* and *attR*, contain a 3-bp long consensus sequence in the core region (5´-TTG-3´) flanked by imperfect repeats (Rausch and Lehmann, 1991; Kuhstoss and Rao, 1991), depicted in Figure 1.3.

Introduction

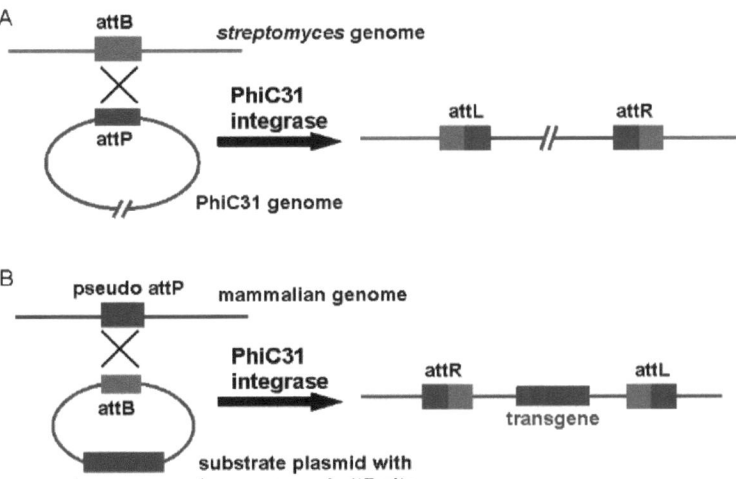

Figure 1.2. Integration mediated by PhiC31 integrase.
(A) Natural integration of the phage genome into the bacterial genome. The phage genome with the attachment site *attP* is shown as an open circle. The upper line represents the bacterial genome, with the bacterial attachment site *attB* (top square). In the presence of PhiC31 integrase the integration reaction proceeds resulting in the linearised phage genome, flanked by two hybrid att sites, *attL* and *attR*. Each hybrid site consists of half attB and half attP sites.
(B) Integration of plasmid DNA into the mammalian genome. The plasmid containing the transgene (lower circle) and the *attB* site is recombined by active PhiC31 integrase into the *pseudo attP* site (top line) of the mammalian genome. After integration into the mammalian genome hybrid *attR* and *attL* sites flanking the integrated transgene are generated. The figure is derived from Calos, 2006.

Attachment site	Sequence 5´- 3´	Length
consensus sequence of pseudo attP	CCTTGGTTAACCTTTAGGTTATCCATGG A AG TAATTGTCAAG CC CAT CG	
attB	TGCGG**GTGCCAGGGCGTGCCC**TTG**GGCTCCCCGGGCGCG**TACTC	34 bp
attP	**GTGCCCCAACTGGGGTAACC**TTT**GAGTTCTCTCAGTTGGGGG**CG	39 bp
attL	TGCGG**GTGCCAGGGCGTGCCC**TTGAGTTCTCTCAGTTGGGGGCG	37 bp
attR	**GTGCCCCAACTGGGGTAACC**TTTGGGCTCCCCGGGCGCGTACTC	36 bp

Figure 1.3. PhiC31 integrase-specific attachment sequences.
The multilevel consensus sequence of pseudo *attP* sites within mammalian genomes (Chalberg et al., 2006), the native *attB* and *attP* sequences and the hybrid sites, *attL* and *attR* are shown. The minimal recognition sequences are highlighted in bold for *attB* (34 bp) with 44 % homology to *attP* and for *attP* (39 bp). The crossover site or overlap region TTG of each attachment site is underlined. The arrows indicate the positions of inverted repeats in *attB* and *attP*. Underlined bases in the *attB* sequence are identical with bases in the wt *attP* site. The hybrid sites *attL* and *attR* are formed by crossover of the sequences mediated by recombination of PhiC31 integrase. Sequences are derived from Thorpe et al., 2000.

With the discovery that SSRs as *Cre* are able to perform recombination in human cells transient assays have demonstrated that preinserted *pseudo*-loxP target sites support Cre-mediated integration and excision in human cell environment (Thyagarajan et al., 2000).

13

Introduction

PhiC31 integrase-mediated recombination efficiency was also successfully tested in intramolecular recombination assays in the presence of *attB* and *attP* recognition sites. The PhiC31 integrase mediated recombination between the full-length *attB* and *attP* sites in the pBCPB+ plasmid vector was determined at a frequency of 52.4 % in human embryonic kidney (HEK)-293 cells, (or only 293 cells), whereas recombination frequency in *E.coli* was 100 %. The minimal sequence length of the integrase specific attachment sites *attB* and *attP* with 34 bp and 39 bp were determined in this assay as well (Groth et al., 2000).

The integration frequency of *attB* site containing plasmid into genomic *pseudo attP* sites of 293 cells was found to be approximately tenfold higher than the frequency of random integration (Thyagarajan et al., 2001). Upon PhiC31 integrase mediated integration of a reporter plasmid encoding a luciferase gene and the *attB* site, stable long-term expression of the reporter gene was shown until four weeks post transfection (Thyagarajan et al., 2001). The integration activity of PhiC31 integrase in liver cells could also be verified in an *in vivo* approach with human blood coagulation Factor IX (hFIX) as reporter. High-pressure tail vein injection of integrase encoding plasmid and hFIX encoding plasmid could show an increase of hFIX levels more than tenfold compared to the control group (inactive integrase) (Olivares et al., 2002). In another study, transgene hFIX expression levels of the active PhiC31 integrase were detected fivefold higher compared to the control group at day 122 post injections even after three times cell cycle inductions (Ehrhardt et al., 2007).

1.4.2 Integration specificity and aberrant integration events

PhiC31 integrase performs unidirectional site-specific integration of a donor plasmid bearing the *attB* site into endogenous randomly scattered attachment sites termed *pseudo attP* sites within the genome. Imperfect *pseudo attP* sites are present in many eukaryotic genomes and have been targeted as insertion sites for transgenes in a site-specific manner in a variety of species. This site-specificity is possible due to a certain degree of homology between the native *attB* site within the donor plasmid and the *pseudo attP* sites present within the human genome. Site-specific recombination into *pseudo attP* sites within human cell lines has been primarily observed in human 293 cells and mouse 3T3 cells (Thyagarajan et al., 2001). Previously inserted *attP* sites present within plasmids, being stable inserted into the cells are competing with *pseudo attP* sites to perform recombination. The analysis of 96 integration events revealed an integration specificity of 15 % (14 out of 96) at which the plasmid was integrated into the previously inserted *attP* site. The characterisation of *pseudo attP* sequences resulted in a sequence homology to the native *attP* site between 20 and 60 %. This implies that PhiC31 integrase does not need perfect sequence fidelity for recognition and recombination between these *pseudo att* sites (Thyagarajan et al., 2001).

With a limited number of rescued integration sites (about 370 *pseudo attP* sites) and a non-random integration profile the PhiC31 integrase system possesses significant advantages over competing integrating systems. The relatively large and complex recognition sequence of about 39 bp seems to possess a low risk of insertional mutagenesis (Chalberg et al., 2006). The same group could show in PhiC31 integration studies, that the total number of rescued sites is lower than integration studies with viral vectors derived from retroviruses and AAVs (Mitchell et al., 2004; Schroeder et al., 2002; Nakai et al., 2003b). More than 500 HIV-1 integration sites were mapped in SupT1 cells revealing a strong target preference for integration in transcription units (70 %) (Schroeder et al., 2002). Although retroviral based integration is not sequence-specific, different retrovirus-derived vectors show various preferences for integration sites in the human chromosomes (Mitchell et al., 2004; Bushman, 2003). PhiC31 integration is also more site-specific compared to SB transposase since integration events upon SB transposition are widely distributed (Yant et al., 2005). A suggested number of more than 10^7 potential integration sites in the genome were assumed as target sites for SB mediated integration (Chalberg et al., 2006).

A detailed study analysing integration specificity in three different eukaryotic cell lines (293 embryonic kidney-derived, HepG2 liver-derived, and D407 retinal pigment epithelium-derived) revealed almost 200 independent integration events in total (Chalberg et al., 2006). The analysis for site-specificity resulted in 56 % integration events. These *pseudo attP* sites are suggested to be recurrent integration sites distributed among 19 "hot spot" *pseudo attP* sequences. The most favoured *pseudo attP* recognition site at location 19q13.31 was targeted with a frequency of 7.5 %. The overall estimation of possibly 370 *pseudo attP* sites confirms the range of wild type (wt) PhiC31 integrase-mediated integration sites between 100 and 1000 as predicted (Chalberg et al., 2006; Thyagarajan et al., 2001). Additional findings highlighted that integrase-specific *pseudo attP* sites are preferentially located in genes or gene dense regions, rather than in promoter regions, likely due to chromatin context effects (Ehrhardt et al., 2006; Chalberg et al., 2006). Integration specificity of PhiC31 integrase was presumably found to be cell line dependent (Chalberg et al., 2006; Ehrhardt et al., 2006; Aneja et al., 2007). It was speculated that the chromosomal context had some influence on specificity (Calos, 2006).

Undesired side effects upon PhiC31 integrase-mediated recombination were first discussed due to chromosomal translocations in primary human fibroblasts being stably

Introduction

transfected with a PhiC31 integrase encoding plasmid (Liu et al., 2006). Recent studies have confirmed integrase-mediated chromosomal abnormalities in primary human embryonic and adult fibroblasts and in mammalian cell lines (Liu et al., 2009). Microdeletions upon recombination (Thyagarajan et al., 2001) and larger deletions of up to 160 bp were already found in mouse liver (Ehrhardt et al., 2005). A large study analysing chromosomal rearrangements in mammalian cells after PhiC31 integrase-mediated integration confirmed aberrations in human cells. Translocations were found linking two different chromosomes upon recombination at approximately 15 % probability besides rarely obtained larger deletions (Ehrhardt et al., 2006). These rearrangements have only been shown in immortalised and rapidly dividing cells *in vitro*, but not yet *in vivo*.

1.4.3 Domain organisation and putative protein structure of the PhiC31 integrase

The coding sequence of PhiC31 integrase encompasses 605 amino acids (AA) corresponding to an open reading frame (ORF) of 1815 nucleotides (Kuhstoss and Rao, 1991; Rausch and Lehmann, 1991). The respective protein has a size of about 67 kDa (Thorpe and Smith, 1998). A second, longer version of 613 AA including an N-terminal addition of an upstream start codon and consists of eight additional AA to ensure full function of the protein (Keravala et al., 2009). This extended version of the integrase is used in the present study.

PhiC31 integrase consists of two large domains, the N-terminal catalytic domain and the C-terminal DNA binding domain, which are connected by a conserved helix (Figure 1.4.B). The large serine recombinases share a highly conserved catalytic serine residue close to the N-terminal end of the enzyme and a similar N-terminal domain with the resolvase/invertase enzymes (Figure 1.4), providing protein-protein interaction and catalysis (Smith and Thorpe, 2002). The function of the relatively large C-terminal domain of large serine integrases within recombination is not yet investigated in detail; however, the C-terminal domain plays an important role in recognition of the attachment sites (Smith and Thorpe, 2002). The same group could recently show a coiled-coil motif with typical heptad repeats comprising two α helices controlling the formation of the synaptic interface in integration and excision. This sequence has been claimed a hyperactive region due to hyperactive mutants found within this motif (McEwan et al., 2009; Rowley et al., 2008). Within this region, the tetramer RFGK comprising amino acid position 451-454 was identified to be crucial for the catalytic activity and responsible for the interaction with the ubiquitary protein DAXX (Chen et al., 2006). The presence of a helix-turn-helix motif within the small C-terminal DNA binding domain (Yuan et al., 2008) suggests putative motifs

Introduction

conferring similar functions within the C-terminal domain of the large recombinase family as well, which serve as DNA-binding elements, such as zinc finger motifs or helix structures.

The domain organisations of the well-characterised γδ resolvase as the paradigm for the resolvase/invertase family and PhiC31 integrase are illustrated (Figure 1.4). Both NTDs show 20 % homology (Rowley and Smith, 2008).

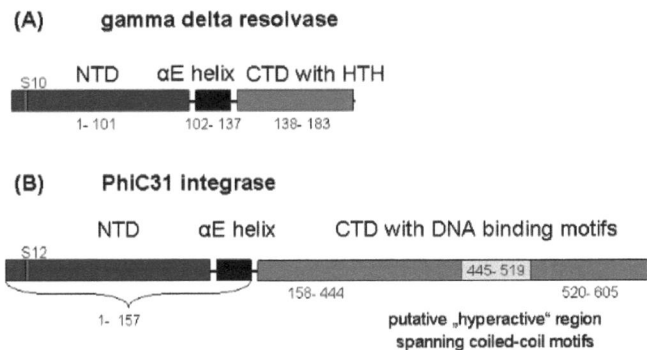

Figure 1.4. Domain organisation of gamma delta resolvase and PhiC31 integrase.
(A) The domain organisation of the γδ resolvase is shown. A larger N-terminal domain (NTD) (left box) is linked via a conserved alpha (α) E helix with the C-terminal domain (CTD) (right box) comprising a helix-turn-helix (HTH) motif. (B) The putative domain organisation of the PhiC31 integrase is illustrated. The NTD provides catalytic activity. The CTD contains DNA-binding motifs and is therefore termed DNA binding domain. A coiled-coil region (445-519) as a structural motif within the secondary structure was recently published (Rowley et al., 2008) according to the Jpred secondary structure prediction server (Cuff and Barton, 2000). The conserved serine residue, which nucleophilically attacks the DNA at the crossover site, is shown at the N-terminus at position 10 or 12 for γδ resolvase or PhiC31 integrase, respectively.

The solved crystal structure of the γδ resolvase (Yang and Steitz, 1995) supports the domain distribution and serves as a putative model for the related enzyme PhiC31 integrase. The crystal structure of the PhiC31 integrase has not been solved to date. The Jpred secondary structure prediction server (Cole et al., 2008; Cuff and Barton, 2000) including COILS prediction (Lupas, 1997) has been used for analysis of the PhiC31 integrase to identify helical regions, extended types of secondary structure and coiled-coil regions (Figure 1.5). Identification of present secondary structure motifs within the protein context might shed some light on the function and the domain structure of the unknown C-terminal DNA binding domain. This might gain a better understanding of particular regions and the effect of mutagenesis at critical residues, which are examined in this thesis. The output of the secondary structure prediction using Jpred 3 [2] is shown in Figure 1.5 below.

---

[2] http://www.compbio.dundee.ac.uk/www-jpred/advanced.html

Introduction

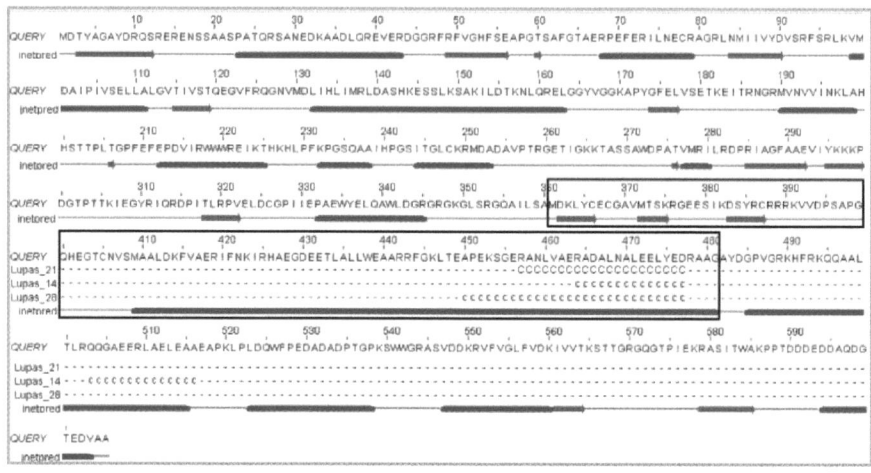

Figure 1.5. Secondary structure prediction of the PhiC31 integrase protein.
The output from Jpred prediction (jnetpred: final secondary structure prediction, Lupas: coil prediction) is shown below the PhiC31 integrase sequence (Cole et al., 2008; Lupas, 1997), illustrated by Jalview a multiple alignment editor (Clamp et al., 2004). The arrows represent extended types of secondary structure as β-sheets. The bars represent helical types as α helices of secondary structure. The coiled-coil helical bundles are a structural motif only found within the large helical structures between amino acid position 450 and 516 using 3 different algorithms (Lupas 21, 14, and 28) according to amino acid length of the particular heptads. Since the coiled-coil motif is only present at the given AA sequence, the Lupas coiled-coil predictions are only shown in the lines where they are present. The putative DNA binding domain encompasses amino acid position 360- 480 and is framed.

The crystal structure of the model enzyme within the serine integrase family has first been solved for the catalytic domain only (Rice and Steitz, 1994) revealing that a formed dimer binds to one DNA strand. The complexed structure of a single resolvase dimer to uncleaved site I DNA and synaptic γδ resolvase tetramer structures bound to two cleaved site I DNA target sequences have also been solved (Yang and Steitz, 1995; Li et al., 2005). These structures were among the available models obtained from the Phyre server (Kelley and Sternberg, 2009) when searching for PhiC31 integrase related similar proteins[3] showing 15 % and 18 % AA identity, respectively. Certain sequence divergence between the PhiC31 integrase and the gamma delta resolvase (Thorpe and Smith, 1998; Yang and Steitz, 1995; Li et al., 2005) exist. However, primary structures are more conserved than corresponding primary sequences (Chothia and Lesk, 1986). This

---

[3] http://www.sbg.bio.ic.ac.uk/phyre/qphyre_output/aa7792bd757b71d0/summary.html

Introduction

assumes similar function within the domains. Different images of the γδ resolvase protein structure are shown in Figure 1.6.

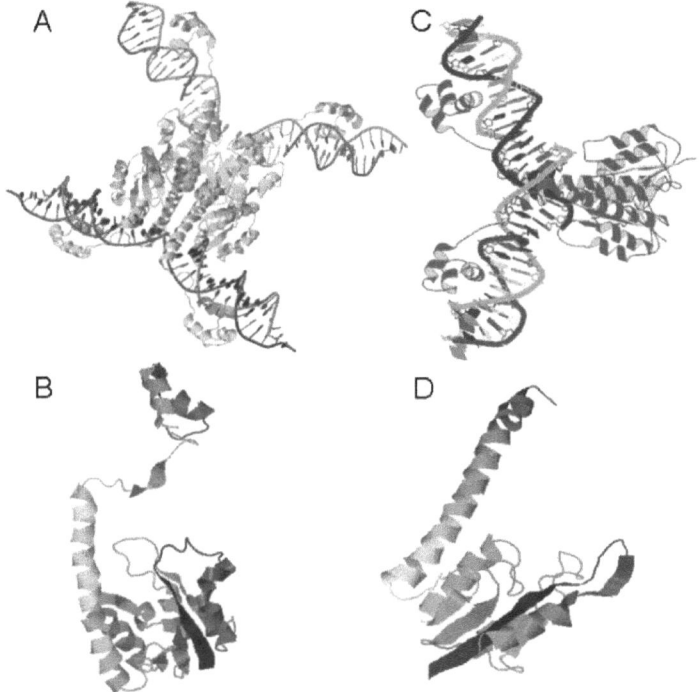

Figure 1.6. Protein structures of the gamma delta (γδ) resolvase.
Two different crystal structures of the γδ resolvase are illustrated. The upper images show the enzyme in its active form complexed with DNA. The lower images show the uncomplexed monomers. The N-terminal end (5´-) is represented in the middle as beta sheets, the C-terminal end is shown shown on top in B and D. (A) The structure of a synaptic γδ tetramer covalently linked to two cleaved DNAs is shown (Li et al., 2005). The dimers are connected to the attachment sites as double strands in the upper left and lower right area. Two DNA double helices are shown, which are cleaved in the middle. The C-terminal end (indicated in Figure B) is shown on the outside relative to the DNA. The protein consists of 183 amino acids. The PDB (Protein data bank) ID is 1ZR2. (B) The gamma delta monomer without DNA from 1ZR2 is shown. (C) The crystal structure of the γδ resolvase complexed with a cleavage site is shown (Yang and Steitz, 1995). The dimer composed of 2 chains is shown and complexed to the DNA (lower strand and upper strand). The protein consists of 140 amino acids. The PDB ID is 1GDTA and 1GDTB. (D) The monomer without DNA from 1GDTA is shown. Protein figures are illustrated at the RCSB PDB website[4] using Jmol version 11.6.

---

[4] http://www.rcsb.org/pdb/home/home.do

Introduction

1.4.4 Applications for PhiC31 integrase

One of the first studies using PhiC31 integrase-mediated gene integration in mice demonstrated delivery *in vivo* by the newly developed hydrodynamic tail vein injection technique of naked DNA to mouse hepatocytes. Therapeutic levels of hFIX were still detectable 250 days after injection due to stable integration and long-term expression even after two-third partial hepatectomy at day 100 post injection (Olivares et al., 2002). In another study, PhiC31 integrase was shown to stably introduce genes in an *ex vivo* approach in primary human keratinocytes grafted onto mice (Ortiz-Urda et al., 2002). A comparative liver study matched the *in vivo* and *in vitro* efficiency and the persistence of PhiC31 integrase with the SB transposon in mouse liver using the hFIX gene as a marker for gene delivery and expression (Ehrhardt et al., 2005). Both systems were capable of producing prolonged levels of hFIX. A novel two-vector system described a hybrid vector that combines stable transduction of the gene-deleted AdV vector and a specific integration potential of the PhiC31 integrase (Ehrhardt et al., 2007).

Since PhiC31 integrase-mediated recombination has been developed to function in mammalian cells, several disease-related studies have been performed to evaluate and improve PhiC31 integrase-mediated gene transfer *in vitro* and *in vivo* (reviewed by Calos, 2006). The PhiC31 integrase has been successfully applied in numerous preclinical gene therapy trials (Olivares et al., 2002; Ortiz-Urda et al., 2002; Hollis et al., 2003; Ehrhardt et al., 2005; Ehrhardt et al., 2007).

Moreover, the PhiC31 integrase system represents a platform technology with additional applications (reviewed by Calos, 2006). Genetic engineering of eukaryotic genomes, production of recombinant proteins and transgenesis have been performed with some success (Allen and Weeks, 2009; Thyagarajan et al., 2008; Sharma et al., 2008; Blaas et al., 2007; Raymond and Soriano, 2007; Thyagarajan and Calos, 2005; Belteki et al., 2003; Hollis et al., 2003; Kolb, 2002). Besides the recombinase-mediated integration, the excision has been extensively investigated within the present study. Both recombination techniques are used for different approaches.

1.4.5 Attempts to improve PhiC31 integrase integration efficacy

Shortly after the PhiC31 integrase has been developed for integration into mammalian genomes, experiments involving mutagenesis by directed evolution via DNA shuffling resulted in enhanced mutant derivatives. Improvements in integration frequency and sequence specificity at a formerly found and preferentially targeted *pseudo attP* site on chromosome 8 were obtained after mutagenesis-selection by DNA shuffling and bacterial screening (Sclimenti et al., 2001). Further attempts to improve the integration efficiency of the PhiC31 integrase were performed using alanine-scanning mutagenesis of the catalytic domain, error-prone PCR or bacterial mutator strains. The chromosomal integration efficiency in cultured human cells could be increased about twofold at both a preintegrated *attP* site and at endogenous *pseudo attP* sites compared to wt PhiC31 integrase (Keravala et al., 2009). In a novel deletion assay a 1.74-fold enhancement of the relative beta-galactosidase activity was found upon C-terminal addition of a nuclear localization sequence (NLS) to the PhiC31 integrase sequence compared to the wt integrase (Andreas et al., 2002).

The generation of fusion proteins by linking the PhiC31 integrase with the hormone binding domain hRP891 gained new techniques for tightly steroid-inducible non-viral integration systems. Sharma et al. (2008) could demonstrate that the drug-regulated PhiC31 fusion proteins increased site-directed gene insertion in the presence of the hormone fourteenfold (Sharma et al., 2008). Recombination activity of a *de novo* synthesised codon-optimised derivative of the native PhiC31 integrase gene was assessed with beta-galactosidase as a reporter gene in embryonic stem cells. Codon-optimisation avoids gene silencing by reducing the number of CpG dinucleotides. This codon-optimised version tagged with the C-terminal NLS of the SV40 large T antigen showed with a twentyfold improvement compared to the wt integrase a similar recombination activity as Cre at inserted minimal target recognition sites flanking the reporter gene in the ROSA 26 based reporter cell lines (Raymond and Soriano, 2007).

Recent attempts beyond improving the enzymatic and the recombination reaction involve improvements of the overall PhiC31 integrase system targeting the expression plasmid and the donor plasmid in order to attenuate and prevent postintegrative gene silencing. This cellular mechanism has been addressed for the first time with the PhiC31 integrase system. Aimed at reducing steady decline in expression, modifications to cellular promoters, plasmid backbone, polyA signals and the addition of an NLS were carefully examined in pulmonary type II cells (Aneja et al., 2009).

Introduction

1.5 Aims of this study

PhiC31 integrase-mediated recombination of plasmid DNA has frequently been applied within the context of therapeutic gene transfer and transgenesis. Improving integration efficiency of recombinases is of critical importance, since the native, preferentially non-viral vector delivery represents the most crucial bottleneck. Specific integration is of general interest since chromosomal aberrations and the risk of insertional mutagenesis demonstrate major drawbacks of the PhiC31 integrase system. The final goal was to improve the recombination activity of PhiC31 integrase within the context of mammalian cells. Site-directed mutagenesis within the integrase DNA-binding domain was performed with the objective of identifying integrase mutants with improved recombination activity. The following was implemented in this study:

1. Screening of integrase mutants showing improved integration efficiency in selection based integration assays in various mammalian cell lines including double mutants and dose dependent studies of the integrase encoding plasmid
2. Screening of integrase mutants revealing enhanced intramolecular excision activity within chromosomal context and within plasmid DNA by means of reporter assays (GFP/luciferase) being valuable in chromosomal engineering as transgenesis and gene activation
3. Evaluation of integrase specificity of selected integrase mutants by detailed analysis of chromosomal insertion sites (*pseudo attP* sites)
4. *In vivo* evaluation of PhiC31 integrase mediated integration efficiency of a substrate plasmid encoding human coagulation factor IX (hFIX) relative to hFIX long-term expression

2. Material

2.1 Laboratory equipment

Bench top centrifuge	Biofuge fresco, Heraeus
Centrifuges (for bacterial cultures)	Sorvall and Eppendorf
Cell-culture centrifuge	Rotanta 460, Hettich
Electroporator	Gene Pulser II, Bio-rad
Heating block	Eppendorf
Incubator for cell culture	Thermo Electron
Light Cycler	Light Cycler 2.0, Roche
Luminometer, Microlumat Plus LB 96V	Berthold Technology
Microwave	Samsung
PCR machine	T professional basic, Biometra
Spectrophotometer	Ultrospec 3000, Pharmacia
FACS Canto	BD
Counting cell chamber	Neubauer
Shaker	Bachofer
Spectrophotometer Ultrospec 2000	Pharmacia Biotech

2.2 Chemicals

Chemicals were purchased from the companies listed below. Supplies in form of sterile plastic ware were either delivered from Peske or Falcon.

Chemicals used and their suppliers

Agar-agar	Roth
Ampicillin	Roth
Amphotericin B	PAA Laboratories
Bacto tryptone	BD Bioscience
Bacto yeast extract	BD Bioscience
BSA (Bovine serum albumine)	Roth
Calcium chloride ($CaCl_2$)	Merck
Carbon tetrachloride (CCl_4)	Riedel-de Haën
Dimethylsulfoxide (DMSO)	Roth

Material	
EDTA (N,N,N,N-Ethylenediaminetetraacetic acid)	Merck
Ethanol (100 %)	Roth
Ethidium bromide	Merck
Formaldehyde	Merck
Glacial acetic acid	Roth
G418 sulphate	PAA Laboratories
Glycerin	Invitrogen
Isopropanol	Invitrogen
Kanamycin	Roth
L-Glutamine	PAA Laboratories
Magnesiumchloride	Merck
Methylene blue	Sigma-Aldrich
Mineral oil	Sigma
Phenol: Chloroform: Isoamyl Alcohol (25: 25: 1) (PCI)	Invitrogen
RNase inhibitor	New England Biolabs (NEB)
SIGMAFAST OPD (o-Phenylenediamine dihydrochloride)	SIGMA
Sodium bicarbonate (7.5 % solution)	Gibco
Sodium chloride	Roth
Tris	Roth
Tris/ HCl	Merck
Trypan blue 0.5 % (w/v)	Biochrom AG
Trypsin	Gibco
Tween 20	SIGMA

2.3 Enzymes

2.3.1 Restriction endonucleases
All restriction endonucleases were ordered from NEB.

2.3.2 Other enzymes

Calf Intestinal Alkaline Phosphatase (CIP)	NEB
KOD polymerase	Novagen
Pfx polymerase	Invitrogen

Material

Polynucleotide kinase (PNK)	NEB
Proteinase K	Merck
Ribonuclease A (RNaseA)	Qiagen
T4 DNA ligase	NEB
Taq polymerase	NEB or Invitrogen

2.4 Solutions, media and buffers

For plasmid isolation

Bacterial resuspension buffer (buffer P1)	50 mM Tris-HCl (pH 7.5) 10 mM EDTA (pH 8.0) 100 µg/ml RNase A
Lysis buffer (buffer P2)	0.2 M NaOH 1 % SDS
Neutralization buffer (buffer P3)	4.09 M guanidine hydrochloride (pH 4.8) 759 mM potassium acetate 2.12 M glacial acetic acid
Lysis buffer for eukaryotic DNA	10 mM Tris 10 mM EDTA 0.5 % SDS

For gel electrophoresis

Solution for agarose gels	1 % agarose in 1× TAE 100 ml 1× TAE buffer
Ethidium bromide staining solution	(10 µg/ml) Ethidiumbromide in 100 ml 1× TAE and agarose heated to 56 °C
50× TAE-buffer	2 M Tris/HCl, pH 8.2 1 M glacial acetic acid 0.1 M EDTA, pH 7.6

Material

10× loading buffer	125 mg (0.25 % w/v Bromphenol blue)
	25 ml Glycerol, 10 % (v/v)
	5 ml EDTA (0.5 M, pH 8.0)
	2.5 ml Tris (1 M, pH 7.5)
1 kb DNA marker	10 µl 1 kb DNA marker (Peqlab)
	20 µl loading buffer (Peqlab)
	70 µl H_2O

For methylene blue staining

Fixing solution	12 ml 37 % formaldehyde
	188 ml D-PBS
Methylene blue solution	2 % methylene blue (Sigma-Aldrich)
	200 ml 50 % ethanol

For incubation of bacteria

LB medium	1 % (w/v) peptone
	0.5 % (w/v) yeast extract
	0.5 % (w/v) NaCl; autoclave
LB Agar	1.5 % agar-agar suspended in LB medium
SOC medium	2 % (w/v) bacto-tryptone
	0.5 % (w/v) bacto-yeast extract
	10 mM NaCl
	2.5 mM KCl
	10 mM $MgCl_2$
	20 mM glucose
	pH to 7.0 and autoclave

For preparation of competent cells

$MgCl_2$ solution	100 mM in H_2O
$CaCl_2$ solution	100 mM in H_2O
	85 mM in 15 % glycerol

For selection of bacteria with antibiotics
Ampicillin solution   50 µg/ml in H_2O

Kanamycin solution   20 µg/ml in H_2O

For ELISA
Coating buffer   0.1 M $NaHCO_3$, pH 9.4

Tris buffered saline (TBS)   10 mM Tris-HCl
                                        150 mM NaCl, in H_2O

TBS-Tween 20 (T)   10 mM Tris-HCl
                                        150 mM NaCl, in H_2O
                                        0.05 % Tween 20

Dilution buffer   10 mM Tris-HCl
                                        150 mM NaCl, in H_2O
                                        5 % BSA
                                        0.05 % Tween 20

2.4.1 Media, supplements and reagents for eukaryotic cell lines

HEK 293 and HeLa cell lines   Dulbecco´s modified Eagle Medium
                                        (DMEM) (Gibco/PAA)
                                        FBS (foetal bovine serum) 10 % (v/v)
                                        (Gibco)
                                        Penicillin/Streptomycin 1 % (v/v) (Gibco)
                                        Amphotericin B (250 µg/ml) 0.1 % (v/v)
                                        (PAA laboratories)

Huh7 and Hep1A cell lines   DMEM
                                        FBS (foetal bovine serum) 10 % (v/v)
                                        Penicillin/Streptomycin 1 % (v/v)
                                        Amphotericin B (250µg/ml) 0.1 % (v/v)
                                        non essential amino acids (100×) (PAA)

Material

HCT cell line

McCoy's 5A medium (GIBCO)
FBS 10 % (v/v)
Penicillin/ Streptomycin 1 % (v/v)
Amphotericin B (250 µg/ml)
sodium bicarbonate 7.5 % (v/v) (GIBCO)

OptiMEM® (for serum-free transfection)   Invitrogen
Dulbecco's phosphate buffered saline   PAA
(D-PBS) (for washing cells)

Freezing media for eukaryotic cells

DMEM 80 % (v/v)
FBS 10 % (v/v)
DMSO 10 % (v/v) (Roth)

G418 solution   50 mg G418 sulphate in 1 ml H_2O
Trypan blue (for cell counting)   0.5 % (w/v) (Biochrom AG)

Transfection reagents for transfection of cell lines
FuGENE6 Transfection Reagent (Roche)
FuGENE® HD Transfection Reagent (Roche)
Lipofectamine 2000 (Invitrogen)

2.5 Kits

ALT Kit (Randox)
Dual Luciferase reporter assay (Promega)
High Pure PCR Product Purification Kit (Roche)
LightCycler FastStart DNA Master^{Plus} SYBR Green I
Nucleobond AX Anion exchange column for quick purification of nucleic acids in large-scale (MACHEREY-NAGEL)
PCR Cloning Kit (Qiagen)
ProtoScript® First Strand cDNA Synthesis kit (NEB)
Pure Yield™ Plasmid Midiprep System (Promega)
Qiagen PCR Cloning kit
QIAquick gel extraction kit (QIAGEN)

QIA midi kit (QIAGEN)
Zero Blunt® TOPO® PCR Cloning Kit (Invitrogen)

2.6 Organisms

2.6.1 Bacteria

For cloning of recombinant DNA and amplification of plasmids the *E.coli* strains DH5α and DH10B were used. For subcloning of PCR products, chemical competent TOP10 bacterial cells (included in pTOPO Cloning Kit) were purchased from Invitrogen. All strains used in this study are summarised in Table 2.1.

Table 2.1. Bacterial strains.

Strain Designation	Genotype
E.coli DH5α	*sup*E44, Δ(lacZYargF)U169(φ80d*lacZ*Δ*M15*), *hsd*R17, *rec*A1, *end*A1, *gyr*A, *thi*-1, *rel*A1; Invitrogen
E.coli DH10B	F- *ara*D139, Δ(*ara, leu*)7697, Δ*lac*X74, *gal*U, *gal*K, mcrA, Δ(mrr-, hsdR, mcrBC), rspL, deoR, (φ80d*lacZ*Δ*M15*), endA1, nup5, recA1; Invitrogen
E.coli Stabl-2	F- endA1, glnV44, thi-1, gyrA96, relA1, Δ(lac-proAB), mcrA, Δ(mcrBC-hsdR-mrr)λ⁻ recA1; Invitrogen
TOP 10	F- mcrA, Δ(mrr-hsdRMS-mcrBC), Φ80lacZΔM15, ΔlacX74 recA1 araD139, Δ(araleu), 7697 galU, galK, rpsL (StrR), endA1, nupG; Invitrogen

2.6.2 Eukaryotic cells

The following cell lines which were used or constructed in this study are depicted in Table 2.2

Table 2.2. Eukaryotic cell lines.

Cell line	Tissue source	Reference
293 (HEK-293)	human embryonic kidney	DSMZ*, Germany
293+*attP*-BGHpolyA-*attB*-eGFP	human embryonic kidney	This study
HeLa	epithelial tissue from human cervical carcinoma	Mark Kay's Laboratory, Stanford University, USA
HCT	human colon carcinoma	Mark Kay's Laboratory, Stanford University, USA
Huh7	human hepatoma cell line	Mark Kay's Laboratory, Stanford University, USA
Hep1A	Mouse hepatoma cell line	Mark Kay's Laboratory, Stanford University, USA
HCT+ p7 (neo^R cell lines upon integrase mediated integration)	human colon carcinoma	This study

Material

2.7 Oligonucleotides

All oligonucleotides used in this study are listed below. Oligonucleotides were obtained from Operon and stored as 10 µM stock solutions. Restriction recognition sites are underlined and respective endonucleases are listed in the right column. Oligonucleotides listed below in Table 2.3 were used to design site-directed mutations in the integrase DNA sequence.

Table 2.3. Oligonucleotides used for site-directed mutagenesis of the PhiC31 integrase.

Name of oligonucleotide and description	Sequence from 5´- to -3´	Restriction recognition
BamHIforwouterNEW	CGG<u>GGATCC</u>GGGTGTCTCGCTACG	*BamHI*
BstEIIrcouterNEW	TAG<u>GGTTACC</u>GCATTCAGCGCGAC	*BstEII*
sdM-K367Aforward	GCCATGGACGCGCTGTACTGC	
sdM-K367Areverse	GCAGTACAGCGCGTCCATGGC	
sdM-E371Aforward	CTGTACTGCGCGTGTGGCGCC	
sdM-E371Areverse	GGCGCCACACGCGCAGTACAG	
sdM-R380Aforward	ACTTCGAAGGCCGGGGAAGAA	
sdM-R380Areverse	TTCTTCCCCGGCCTTCGAAGT	
sdM-K386Aforward	GAATCGATCGCGGACTCTTAC	
sdM-K386Areverse	GTAAGAGTCCGCGATCGATTC	
sdM-R394Aforward	TGCCGTCGGCGAAGGTCGTC	
sdM-R394Areverse	GACCACCTTCGCGCGACGGCA	
sdM-D398Aforward	AAGGTGGTCGCCCCGTCCGCA	
sdM-D398Areverse	TGCGGACGGGGCGACCACCTT	
sdM-D417Aforward	GCGGCACTCGCCAAGTTCGTT	
sdM-D417Areverse	AACGAACTTGGCGAGTGCCGC	
sdM-R423Aforward	GTTGCGGAAGCCATCTTCAAC	
sdM-R423Areverse	GTTGAAGATGGCTTCCGCAAC	
sdM-E435Aforward	GAAGGCGACGCAGAGACGTTG	
sdM-E435Areverse	CAACGTCTCTGCGTCGCCTTC	
sdM-R446Aforward	GAAGCCGCCGCACGCTTCGGC	
sdM-R446Areverse	GCCGAAGCGTGCGGCGGCTTC	
sdM-K457Aforward	GCGCCTGAGGCGAGCGGCGAA	
sdM-K457Areverse	TTCGCCGCTCGCCTCAGGCGC	

Material

Name of oligonucleotide and description	Sequence from 5´- to -3´	Restriction recognition
sdM-D470Aforward	GAGCGCGCCCCGCCCTGAAC	
sdM-D470Areverse	GTTCAGGGCGGCGGCGCGCTC	
sdM-D366Aforward	TGTCCGCCATGGCCAAGTGTACTG	
sdM-D366Areverse	CAGTACAGCTTGGCCATGGCGGACA	
sdM-E382Aforward	CGAAGCGCGGGGCAGAATCGATCAAG	
sdM-E382Areverse	CTTGATCGATTCTGCCCCGCGCTTCG	
sdM-E383Aforward	GCGCGGGGAAGCATCGATCAAGGAC	
sdM-E383Areverse	GTCCTTGATCGATGCTTCCCCGCGC	
sdM-R390Aforward	GGACTCTTACGCCTGCCGTCGCCGG	
sdM-R390Areverse	CCGGCGACGGCAGGCGTAAGAGTCC	
sdM-R393Aforward	CCGCTGCCGTGCCCGGAAGGTGG	
sdM-R393Areverse	CCACCTTCCGGGCACGGCAGCGG	
sdM-E406Aforward	GGGCAGCACGCAGGCACGTGCAAC	
sdM-E406Areverse	GTTGCACGTGCCTGCGTGCGCCC	
sdM-R429Aforward	CAACAAGATCGCGCACGCCGAAGG	
sdM-R429Areverse	CCTTCGGCGTGCGCGATCTTGTTG	
sdM-E4329Aforward	CAGGCACGCCGCAGGCGAC GAAG	
sdM-E4329Areverse	CTTCGTCGCCTGCGGCGTGCCTG	
sdM-K450Aforward	GACGCTTCGGCGCGCTCACTGAGGC	
sdM-K450Areverse	GCCTCAGTGAGCGCGCCGAAGCGTC	
sdM-R461Aforward	GAGCGGCGAAGCGGCGAACCTTGTTG	
sdM-R461Areverse	CAACAAGGTTCGCCGCTTCGCCGCTC	

Oligonucleotides listed below were used for cloning and sequencing (Table 2.4).

Material

Table 2.4. Oligonucleotides used for cloning and sequencing.

Name of oligonucleotide and description	Sequence from 5´- to 3´	Restriction recognition
BamHINotIBstEIIfor	GATCCGCGGCCGCG	*NotI*
BstEIINotIBamHIrev	GTTACCGCGGCCGCG	*NotI*
PmeINcoIchr.2	GTTTAAACCCATGGGGCAGGTGGCG CAGTCAA	*PmeI, NcoI*
Chr.2NcoIPmeI	GTTTAAACCCATGGCCTGGGGCTGAA GTTTGAT	*NcoI, PmeI*
SpeIXhoIHindIIIchr12	GGACTAGTCCCTCGAGGCCCAAGCT TGGGTGGCTCTAGCGTCTACGATG	*SpeI, XhoI, HindIII*
Chr12HindIIISpeI XbaI	TGCTCTAGAGCAGACTAGTCCCAAGC TTGGGTGAGATATGCGGCAAAAACA	*HindIII, SpeI, XbaI*
SpeIEcoRVHindIII chr19	GGACTAGTCCGATATCCCCAAGCTTG GGTTGTTTGGCTCAGACCTTCC	*SpeI, EcoRV, HindIII*
Chr19HindIIIEcoRI SpeI	GGACTAGTCCGATATCCCCAAGCTTG GGTGTTGGTAATTTGCGGTTCA	*HindIII, EcoRV, SpeI*
Luciferaserev.5	CCATCTTCCAGCGGATAGAATGGC	
Int-Pst5`-	CGCCTGCAGGTACCGGTCCGGAAT	*PstI*
Int-BstEII3`-	TGCGGTAACCCTCAATCTTCGTGG	*BstEII*
forwAgeIhs2	ACCGGTAGTGAGCTCCTGTTGCT	*AgeI*
revStuIhs2	AGGCCTTCCTGGGGCTGAAGTTTGAT	*StuI*
forwAgeIhs12	ACCGGTTGTCTCAAGAGGGAAGTG	*AgeI*
revStuIhs12	AGGCCTAAGCTTGGGTGAGATATGC	*StuI*
forwAgeIhs19	ACCGGTGTTGTTTGGCTCAGACCTT	*AgeI*
revStuIhs19	AGGCCTTGTTGGTAATTTGCGGTTCA	*StuI*
attB-F for sequencing	TACCGTCGACGATGTAGGTCACGGTC	
attB-R for sequencing	CGTGACCACCGCGCCCAGCGGTTT	
forw.DAXXqRT	AGGAGTTGGATCTCTCAGAA	
rev.DAXXqRT	TGATGAGCCGCTCAATG	

Duplex RNA listed in Table 2.5 below was used for quantitative real-time PCR in stocks of 100 pmoles/µl and 30 pmoles/µl for DAXX siRNA and GL-siRNA.

Table 2.5. Duplex RNA used for quantitative real-time PCR.

Name of siRNA and description	Sequence from 5´- to 3´
DAXX-siRNA	GGAGUUGGAUCUCUCAGAATT
GL-siRNA (GAPDH specific) as control	CUUACGCUGAGUACUUCGATT

Material

2.8 Plasmids

Plasmids constructed in the context of this study are shown in Table 2.6 and are described in the results section.

Table 2.6. Plasmids used for cloning and transfection.

Plasmid name	Plasmid Description	Reference
pCS Int wild type	Backbone plasmid contains wild type PhiC31 integrase as template plasmid for point mutants, Amp^R,	Olivares et al., 2002, Ehrhardt et al., 2005
mInt	Integration deficient negative control plasmid of pCS Int, contains point mutation at S20F in catalytic centre, Amp^R	Olivares et al., 2002, Ehrhardt et al., 2005
p7 derivative of p11	Substrate plasmid for PhiC31 integrase, contains attB site, high copy origin of replication from Tn5, Kan^R	Ehrhardt et al., 2005
pBSattBhFIX	Substrate plasmid for PhiC31 integrase with attB site, hFIX minigene, liver-specific promoters and enhancer, Amp^R	Ehrhardt et al., 2005
pCS Int + D366A	Integrase gene contains D366A mutation	This study
pCS Int + K367A	Integrase gene contains K367A mutation	This study
pCS Int + E371A	Integrase gene contains E371A mutation	This study
pCS Int + R380A	Integrase gene contains R380A mutation	This study
pCS Int + E382A	Integrase gene contains E382A mutation	This study
pCS Int + E383A	Integrase gene contains E383A mutation	This study
pCS Int + K386A	Integrase gene contains K386A mutation	This study
pCS Int + R390A	Integrase gene contains R390A mutation	This study
pCS Int + R393A	Integrase gene contains R393A mutation	This study
pCS Int + R394A	Integrase gene contains R394A mutation	This study
pCS Int + D398A	Integrase gene contains D398A mutation	This study
pCS Int + E406A	Integrase gene contains E406A mutation	This study
pCS Int + D417A	Integrase gene contains D417A mutation	This study
pCS Int + R423A	Integrase gene contains R423A mutation	This study
pCS Int + R429A	Integrase gene contains R429A mutation	This study
pCS Int + E432A	Integrase gene contains E432A mutation	This study
pCS Int + E435A	Integrase gene contains E435A mutation	This study
pCS Int + R446A	Integrase gene contains R446A mutation	This study
pCS Int + K450A	Integrase gene contains K450A mutation	This study
pCS Int + K457A	Integrase gene contains K457A mutation	This study
pCS Int + R461A	Integrase gene contains R461A mutation	This study
pCS Int + D470A	Integrase gene contains D470A mutation	This study
pLucCR wt	Substrate plasmid for integrase with wt attP & attB site and firefly luciferase, Amp^R	Aneja et al., 2007, Carsten Rudolph's laboratory, Munich
pLuc+hs2	Substrate plasmid for integrase with pseudo attP from chr. 2q11.2, Amp^R	This study
pLuc+hs12	Substrate plasmid for integrase with pseudo attP from chr. 12q22, Amp^R	This study
pLuc+hs19	Substrate plasmid for integrase with	This study

Material

Plasmid name	Plasmid Description	Reference
	pseudo attP from chr. 19q.13.31, Amp^R	
phRL-0	Human codon-optimised *Renilla* luciferase containing vector, Amp^R	Invitrogen
pRL-TK	*Renilla* luciferase vector with TK promoter, Amp^R	Invitrogen
pGL3	contains firefly luciferase gene, SV40 promoter, Amp^R	Invitrogen
pCRBluntIITOPO	Cloning vector for inserts: PCR fragments or restricted fragments, Kan^R	Invitrogen
pHM5	Cloning vector, Kan^R	Marc Kay's laboratory, Stanford University, USA
pEGFP	p*attP*-BGHpolyA-*attB*-EGFP with CMV, eGFP, SV40, Kan^R	Chen et al., 2006
pCRmOrange	Compensation plasmid to evaluate transfection efficiency detected by FACS, Amp^R, Kan^R	Invitrogen
pQCXIX-GFP-I-MDR	eGFP expressing plasmid as control for FACS experiments, Amp^R	Wolfgang Pfützner's laboratory, Munich
pDrive+DAXX	DAXX expressing plasmid as internal control for real-time PCR	This study
pBS	Cloning vector used as stuffer plasmid DNA, Amp^R	Marc Kay's laboratory, Stanford University, USA
pUC19	High copy cloning vector used as stuffer plasmid DNA, Amp^R	Marc Kay's laboratory, Stanford University, USA

3. Methods

3.1 General methods with eukaryotic cell culture

3.1.1 Cultivation of eukaryotic cells

In this work only adherent cell lines were cultured and transfected. All cell lines were grown in 10-cm tissue culture dishes in appropriate medium at 37 °C and 5 % CO_2 with supplements as described in section 2.4.1 until confluent at approximately 90 %. Cells were passaged twice a week. For splitting, cells were washed with 3 ml D-PBS, and 3 ml trypsin (PAA) was added to detach cells from the plastic surface. Cells were centrifuged at 1.500 rotations per minute (rpm) for 3 minutes (min) in a Rotana 460 cell culture centrifuge. The supernatant was discarded, and cells were resuspended in medium and cells were diluted into differently sized culture plates (24-wells plate, 6-wells plate, 6-cm dish or 10-cm dish).

3.1.2 Storage of eukaryotic cells

For long-term storage in liquid nitrogen, cultured cells were treated with trypsin as described above and resuspended in 2 ml freezing medium containing 10 % DMSO as a cryoprotectant. Resuspended cells were aliquotted into two cryo tubes, slowly frozen down at -80 °C and stored in liquid nitrogen until use.

3.1.3 Cell counting of cultured eukaryotic cells

The number of cells was determined with a Neubauer counting cell chamber. Trypsinised cells were diluted with trypan blue at a ratio of 1:1. Blue dye is only taken up by dead cells. Therefore, dead blue stained cells could be distinguished from colourless, living cells. Twelve microlitres of this mix were transferred onto the counting chamber and living cells present within the grid were counted. The total cell number within four large squares was averaged and the dilution factor was taken into consideration in order to determine the exact number of cells. Multiplication of counted cells with factor 10^4 represented the entire cell number per millilitre (ml).

3.1.4 Transfection of eukaryotic cells with plasmid DNA

Mammalian cells were transfected with either FuGENE6, FuGENE® HD (Roche) or Lipofectamine 2000 (Invitrogen). Lipofection or liposome transfection alters the cellular plasma membrane allowing any kind of nucleic acid complexed with liposomes to cross into the cytoplasmic space. Short transfection protocols using either FuGENE® HD and

Methods

FuGENE6 (a) or Lipofectamine 2000 (b) are described below. Additional information is given in respective manuals.

(a) A defined number of cells, depending on growth, size and origin, were seeded in the appropriate well size of the culture dishes the day before transfection to reach a confluency of 60-90 %. According to well size and number of replicates the transfection reagent (here FuGENE) was diluted with serum-free Opti-MEM. After five minutes of incubation, the appropriate number and concentration of plasmid DNA (pDNA) was added to the medium. Total DNA (in ng) and appropriate transfection reagent (in µl) were used at a ratio of 3:1 unless stated otherwise (Table 3.1). The transfection mix was incubated for at least 15 min at room temperature before it was added dropwise to the cells and incubated for two days. For the FuGENE6 system, cells were transfected at low cell densities 50-80 % confluent. For the FuGENE HD system, cells were transfected at high densities ranging between 80 and 90 %.

(b) One day before transfection, cells were plated so that cells would be 90-95 % confluent at the time of transfection. DNA was diluted in appropriate volume of Opti-MEM medium and gently mixed. Lipofectamine 2000 was diluted in the same amount of Opti-MEM medium and incubated for 5 min at room temperature. Upon incubation, diluted DNA and diluted Lipofectamine were combined, gently mixed and incubated at room temperature for 20 minutes. The complex was subsequently added to each well or dish containing cells and medium. Cells were gently mixed by rocking the plates and incubated at 37 °C and 5 % CO_2. Cells were harvested and assayed two days post transfection for transgene expression in respective assays.

Table 3.1. Guidelines for transfection reagents in different well size using FuGENE6 transfection reagent: DNA ratio of 3:1.

Well size format	Seeding density*	Volume of FuGENE6 reagent	Mass of DNA	Total volume of complex	Growth medium
24 well- 1 well	0.05×10^6	1.2 µl	0.4 µg	20 µl	1 ml
6 well- 1 well	0.3×10^6	6.0 µl	2.0 µg	100 µl	3 ml
6-cm dish	0.8×10^6	10 µl	5.0 µg	250 µl	5 ml

*The number of cells in a confluent well varies depending on the cell type. This table shows the approximate number of HeLa cells being used for transfection. For 293 cells, approximately twice as many cells are required since 293 cells are smaller in size.

3.1.5 Transfection with siRNA and plasmid DNA and further applications

HEK 293 cells were seeded into 6-well plates to reach confluency of about 30 % the next day. Transfections were carried out according to the protocol (Invitrogen). One hundred

Methods

picomoles (pmol) siRNA were transfected with 5 µl Lipofectamine 2000 and 250 µl Opti MEM. The cells were incubated at 37 °C and 5 % CO_2. One day later 2000 ng plasmid DNA were additionally transfected with 6 µl Lipofectamine 2000 and 100 µl Opti MEM. Two days post-transfection, the cells were either harvested and prepared for real-time PCR (RT PCR) or were splitted and cultivated under selection for quantification of integration events (Figure 3.1). Transfection setups using siRNA and plasmid DNA for each group are described in Table 3.2.

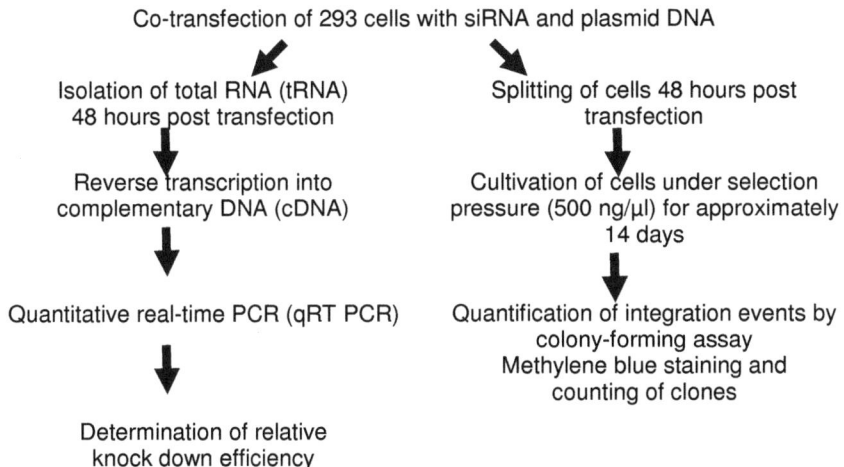

Figure 3.1. Overview of process after with siRNA and plasmid DNA

Table 3.2. Transfection setup and applications.
Transfected HEK 293 cells were divided into different groups based on the amount of plasmid DNA and siRNA used in the transfection. All groups were determined for integration efficiency by a colony-forming assay (CFA). Groups II and III were used in RT PCR.

Group	Transfected Plasmids	Transfected siRNA in pmoles	Further applications
I	1000 ng pCS Int + 1000 ng p7	1000 ng stuffer plasmid as control	CFA
II	1000 ng pCS Int + 1000 ng p7	100 pmol DAXX specific siRNA	CFA, real-time PCR
III	1000 ng pCS Int + 1000 ng p7	100 pmol unspecific siRNA (GAPDH)	CFA, real-time PCR
IV	1000 ng mInt + 1000 ng p7	100 pmol DAXX specific siRNA	CFA
V	1000 ng mInt + 1000 ng p7	100 pmol unspecific siRNA (GAPDH)	CFA

Methods

3.2 Process to select stable cell lines with integrated plasmid

Colony-forming assays were based on co-transfection of substrate plasmid p7 and integrase encoding plasmid into cell lines. The substrate plasmid p7 carrying a neomycin resistance gene and a recognition *attB* site was recombined with *in trans* delivered PhiC31 integrase at *pseudo attP* sites (*attP´*) within the genome resulting in the formation of hybrid *attL/attR* sites. Stable substrate plasmid integration into the genome was achieved under selection pressure in cell culture medium supplemented with 500 ng/ml neomycin. The substrate plasmid p7 and the PhiC31 integrase encoding plasmid were used for co-transfection into cell lines and the plasmid maps are illustrated in Figure 3.2. The co-transfection of the substrate plasmid and integrase encoding plasmid is outlined in Figure 3.3 below. Recombination of the *attB/attP´* sites leading to integration of substrate plasmid p7 into the genome and the selection of stable single cell clones is shown.

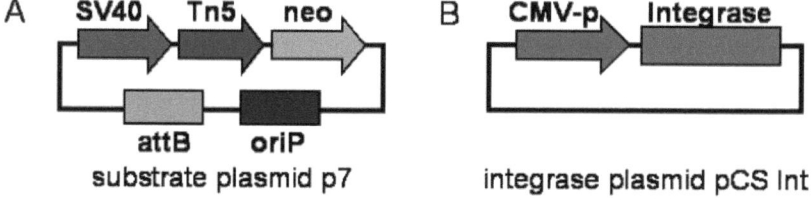

Figure 3.2. The plasmids used for co-transfection into cell lines are shown.
(A) The substrate plasmid p7 contains the simian virus SV40 promoter, the Tn5 promoter, the neomycin-resistance gene (neo), the integrase PhiC31 attachment site *attB* and the origin of replication (oriP). (B) The integrase plasmid pCS Int encodes a cytomegalovirus virus promoter (CMV-p) and an integrase encoding gene, which should represent all types of integrase (wt Int, mInt or designed mutants). The expressed integrase mediates recombination of the substrate plasmid p7 at *attB* × *attP´* within the genome.

```
Co-transfection of cells with integrase expressing plasmid (pCS Int)
and substrate plasmid p7 containing attB site (in 6-well plate format)
                            ↓
    Splitting of cells 48 hours post transfection into 10-cm dishes
                            ↓
Cultivation of cells under selection pressure (500 µg/ml G418) for approximately 14 days
                       ↙         ↘
```

Quantification of   **Establishment of stably transfected**
integration events (3.2.1)   **cell lines (3.2.2)**
(Colony-forming assay)   Isolation of single cell clones and
Methylene blue staining and   amplification
counting of clones
                            ↓
              Isolation of genomic DNA (3.2.3)
                            ↓
              Analysis of integration events
              by plasmid rescue (3.2.4)
- Triple digestion (*SpeI*, *NheI*, and *XbaI*)
- Re-ligation of genomic fragments
- Selection by transformation
- Plasmid purification
- Control digestion (*PstI*)
- Sequencing from hybrid *attL*/*attR* sites
into chromosomal DNA

Figure 3.3. Outline of plasmid transfection, cultivation and subsequent applications. The assays and methods are in bold. The respective section number for each activity is shown in brackets.

3.2.1 Quantification of integration events

PhiC31 integrase mediated integration efficiencies were quantified by a colony-forming assay (CFA). As control groups, cells were transfected with the wild type integrase (wt Int) as positive control or with the mutated integrase (mInt) as negative control. Cells were transfected in triplicate with 2000 ng pDNA/ well and plated on either 6-cm dishes or 6-well plates. Forty-eight hours post-transfection, the cells were counted and split in triplicate into 10-cm dishes at a seeding density of 4×10^4 or 4×10^5 cells/10-cm dish. Twenty-four hours later, the medium was supplemented with 500 µg/ml G418. Stable genomic integration of the substrate plasmid conferred resistance to the transfected cells for G418 by the neomycin resistance gene neo^R encoding an aminoglycoside 3'-phosphotransferase. Medium supplemented with 500 µg/ml G418 was changed every three to four days for

Methods

approximately two weeks keeping the selection pressure maintained until individual cells grew into clearly separated colonies. For the subsequent quantification of integration events, the cultures were stained with methylene blue. Medium was removed and the clones were washed once by adding 3 ml D-PBS onto each 10 cm dish. Cell clones were fixed by addition of 3 ml of 3 % formaldehyde solution and incubation for 10 min. After washing with 3 ml D-PBS, 3 ml methylene blue solution were added to the dishes and the cells were stained at room temperature for 5 min. Methylene blue was removed and plates were rinsed with water and dried. Clones in size of about 2 millimetres and larger were counted and evaluated as integration events.

3.2.2 Establishment of stably transfected cell lines

Twenty-four hours post-transfection, 500 µg/ml G418 were added to the cells cultivated in a 6-well format. Selective medium supplemented with G418 was changed every three to four days for about two to three weeks until colonies were observed as individual cell clones. Single clones were picked and individually isolated (monoclonal) from the original dish either by tips or by means of cloning rings filled with 40 µl trypsin. Trypsinized cells were transferred into one well of a 24-well plate. Cell-clones were then established as new cell lines by amplifying them in selective medium.

3.2.3 Isolation of genomic DNA from eukaryotic cells

Confluent cells grown in a 10-cm dish were treated with 3 ml trypsin. Detached cells were transferred into 1.5 ml Eppendorf tubes and harvested by centrifugation for 2 min at 2000 rpm. The supernatants were discarded, and the cells were washed once with D-PBS. The cells were harvested by centrifugation. Eukaryotic cell pellets were resuspended in 200 µl D-PBS. Two hundred micro litres of lysis buffer were added, and the content was mixed in tubes. Thirty micro litres of 10 % SDS were added followed by 20 µl (20 µl/ml) Proteinase K. The solution was mixed thoroughly and incubated overnight at 55 °C with agitation. The next day 2 µl RNaseA were added. After incubation for 30 min at 37 °C 350 µl phenol-chloroform-isoamylalcohol (PCI) were added to remove proteins. After vortexing briefly, the cells were centrifuged for 2 min at maximum speed in a table top centrifuge. The supernatants were transferred to new Eppendorf tubes and extracted with PCI again. The supernatants were transferred to new Eppendorf tubes. Fifty microlitres NaAc (pH 5.0, 3 M) and 1 ml 100 % ethanol were added to precipitate the DNA. The solution was vortexed until the genomic DNA was visible. The genomic DNA was pelleted by centrifugation at maximum speed for 10 min. The supernatants were discarded and 500 µl of 70 % ethanol were added. Another centrifugation step at maximum speed was performed for 2 min and the genomic DNA samples were washed with 500 µl 70 % ethanol at room temperature with agitation for at least 30 min. The centrifugation step was repeated and genomic DNA pellets were air dried and resuspended in 120 µl TE buffer at 37 °C with agitation until DNA became viscous. Genomic DNA concentrations were estimated with Ultrospec 3000 and stored at -20 °C until needed further.

3.2.4 Analysis of integration events by plasmid rescue

Up to 10 µg of genomic DNA from clones were digested with the restriction enzymes *SpeI, XbaI,* and *NheI* purified, and eluted in 20 µl H_2O. Digested DNA was religated using 20 µl purified restricted genomic DNA, 5 µl 10 × ligation buffer, 1 µl ligase, and H_2O up to 50 µl. The ligation mixture was precipitated with ethanol and resuspended in 20 µl sterile H_2O, 1 µl transformed into DH10B cells by electroporation. Single colonies were cultured and plasmid-isolated. Rescued plasmids were analysed by restriction enzyme digestion with *PstI*. Plasmids with a characteristic fragment size of 1.6 kb and additional fragments were sequenced with specific sequencing primers. The plasmid rescue described in Figure 3.3 is illustrated in Figure 3.4 below with the analytical *PstI* digestion and the sequencing of flanking chromosomal DNA.

Methods

Digestion of chromosomal DNA with SpeI, NheI, and XbaI

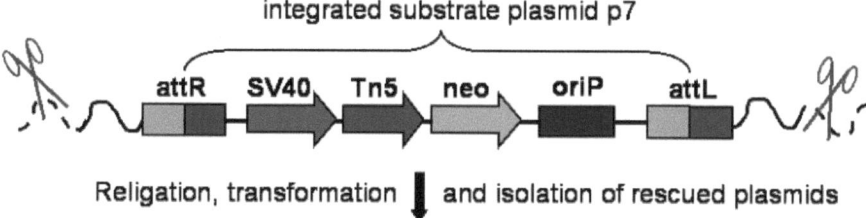

Religation, transformation ↓ and isolation of rescued plasmids

Analytical digestion and sequencing into chromosomal DNA

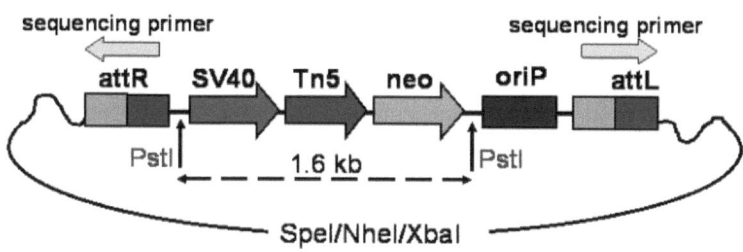

Figure 3.4. The plasmid rescue.
After genomic DNA was isolated from cell lines containing the substrate plasmid p7, a triple digest was performed with *SpeI, NheI* and *XbaI*. Genomic fragments were further re-ligated and transformed into *E.coli*. Isolated plasmids were analysed for integration events by analytical digestion with *PstI* resulting in a specific fragment size of 1.6 kb. Sequencing with *attR/attL* specific sequencing primers could determine the target location, at which the plasmid was recombined into genomic DNA.

3.2.5 Estimation of the quantity of colonies dependent on the integrase plasmid concentration

Several integrase mutants were screened for dose dependency. In these experiments, integrase plasmid amounts were either increased up to twentyfold or decreased down to twentyfold compared to the amount of plasmid p7 transfected. The prerequisite for comparative studies with different amounts of integrase plasmids was that the amount of substrate plasmid p7 was kept constant in all experiments. To ensure that equal amounts of total pDNA were transfected per dish the appropriate amount of stuffer DNA, pBS or pUC19, was added. The plasmid ratios used in the initial assay and in dose dependent studies are listed in Table 3.3. Transfections were carried out as described in Section 3.1.4 (Table 3.1) in 6-well plates.

Table 3.3. Overview over transfection setups with increasing and decreasing integrase plasmid amounts.
Different transfection conditions based on weight ratio and molar ratio used for CFA to analyse integration efficiencies are listed. The initial CFA experiments were performed at a weight ratio of 1:1, which corresponds to a molar ratio of 0.48:1, taking the size of the transfected plasmids into consideration; (pCS Int = 6230 bp, substrate plasmid p7 = 2966 bp). The dose dependent studies with increasing and decreasing amounts of Integrase plasmid were based on molar transfection ratios of substrate plasmid and integrase plasmid.

	Molar ratio for transfections p7: integrase plasmid	Amount of p7 substrate plasmid in ng	Amount of integrase plasmid in ng	Amount of pUC19 stuffer plasmid in ng
initial screening ratio in HeLa cells	1 : 0.48	1000	1000	/
decreasing amounts of integrase plasmid	1 : 0.025	50	2.6	1947.4
	1 : 0.1	50	10.3	1939.7
	1 : 0.5	50	52	1898
increasing amounts of integrase plasmid	1 : 1	50	102	1848
	1 : 5	50	513	1437
	1 : 10	50	950	/
	1 : 20	50	1950	/

3.3 Molecular biology techniques

3.3.1 Strain cultivation and storage of bacteria

E.coli strains were cultured in autoclaved sterile LB medium. For solid agar 1.5 % agar was added. Media were supplemented with 50 µg/ml respective antibiotic ampicillin (Amp) or 20 µg/ml kanamycin (Kan). Bacteria on LB agar were cultivated overnight at 37 °C. Bacteria in liquid culture were cultivated overnight at 37 °C on a shaking platform at 200 rotations per minute (rpm). For long-term storage, bacteria were resuspended in LB medium supplemented with 10 % glycerol and frozen at -80 °C.

3.3.2 Transformation of bacteria

a) Preparation of electro-competent cells and electro-shock transformation of bacteria

Bacteria from DH10B stock solution were incubated on LB only plates overnight at 37 °C. One single colony was used to inoculate 5 ml low salt LB medium at 37 °C overnight with agitation. The culture was used to inoculate 1 litre (l) LB medium at 37 °C until OD_{600} had reached 0.6- 0.7. After cells were chilled on ice for 40 min, centrifugation was performed at 4000 rpm for 10 min at 4 °C. Bacterial pellet was resuspended in 1 l H_2O. Centrifugation steps were repeated twice while cells were resuspended in 500 ml 10 % glycerol. Bacterial

Methods

pellet was resuspended in 20 ml 10 % glycerol, centrifuged and resuspended in two to three ml ice-cold 10 % glycerol, aliquoted in volumes of 50 µl, frozen in liquid nitrogen and stored at -80 °C.

b) Electroporation

Fifty microlitres of competent cells were thawed on ice. One microlitre purified plasmid DNA or up to 2 µl newly ligated DNA was added. The mixture was transferred into a prechilled electroporation cuvette on ice. Cells and DNA were electroporated with a Bio-rad GenePulser II electroporator under the following characteristics: 2.5 kV (kilovolt), 200 Ω (Ohm), 25 µF (microfarad). Immediately after electroporation 600 µl Super Optimal Broth (SOC) medium were added to the cells. The suspension was transferred into a 1.5 ml Eppendorf tube shaken at 37 °C for 1 hour. The appropriate amount of cell suspension was plated onto appropriate solid agar and incubated at 37 °C overnight.

c) Preparation of heat-shock competent cells and heat-shock transformation of bacteria

Chemical competent E.coli cells were made using strain DH5α. Cells were streaked for isolation on LB, incubated at 37 °C overnight. One single colony was used to inoculate 3 ml LB medium which was shaken overnight at 37 °C. Three millilitres culture was used to inoculate 1 l LB medium until OD_{600} reached 0.3-0.4. All centrifugation steps were performed at 5000 rpm for 10 min at 4 °C. After the initial centrifugation, the pellet was resuspended in 100 mM 250 ml $MgCl_2$ (magnesium chloride), the cells were centrifuged, and the pellet was resuspended in 100 mM 50 ml $CaCl_2$. 450 ml 100 mM $CaCl_2$ were added and the mixture was chilled on ice for 20 min. The centrifugation was repeated, and the cells were resuspended in 85 mM 20 ml $CaCl_2$ (diluted in glycerol). Aliquots in one hundred microlitres were immediately frozen down in liquid nitrogen and stored at -80 °C.

d) Heat-shock transformation

One vial of chemically competent cells were thawed on ice. One µl plasmid DNA (pDNA) or two µl newly ligated DNA were added to the cells and chilled on ice for 10 min. The suspension was then incubated at 42 °C for 45 seconds and subsequently transferred on ice for 2 min. After 600 µl of SOC medium were added to the transformed cells, the mix was shaken 1 hour at 37 °C. Cells were then plated onto selective agar and incubated at 37 °C overnight.

Methods

3.3.3 Preparation of plasmid DNA
For plasmid DNA isolation, *E.coli* culture volumes were chosen based to the need for purified plasmid DNA.

a) Plasmid mini preparation

The plasmid mini preparations were prepared from 3 ml liquid bacterial culture. Overnight cultures were transferred into 1.5 ml Eppendorf tubes and centrifuged for 2 min at maximum speed (13000-14000 rpm). Supernatants were discarded and bacterial pellets were resuspended in 100 µl resuspension buffer P1. One hundred µl lysis buffer P2 were added and Eppendorf tubes were turned upside down several times and incubated for three minutes. After adding 100 µl neutralization buffer P3, tubes were subsequently shaken and centrifuged in a table top centrifuge at 13000 rpm for 5 min. Supernatants were then transferred into new Eppendorf tubes. Thirty microlitres NaAc (3M, pH 5) and 600 µl ice-cold 100 % ethanol were added for precipitation. The tubes were vortexed for three seconds and centrifuged at maximum speed for 10 min. Supernatants were discarded, and 70 % ethanol was added. Another centrifugation step at maximum speed for 10 min was performed. The tubes containing the DNA pellets were inverted and briefly dried upside down. The pellets were resuspended in 30-40 µl H_2O or TE buffer.

b) Plasmid preparations in large-scale: Midi and Giga preparations

The Pure Yield™ Plasmid Midiprep System from Promega was used for bacterial culture volumes between 50-100 ml. For pDNA purification in Gigaprep scale (bacterial cultures of 1-2 l) the Nucleobond® AX anion exchange columns for quick purification of nucleic acids from MACHEREY-NAGEL were used. Bacterial cultures (100 ml up to 2 l) supplemented with antibiotic were grown at 37 °C with agitation overnight. The cultures were centrifuged at 8000 rpm for 10 min at room temperature. The pellets were either stored at -20 °C or used immediately for DNA purification according to the manufacturer's instructions. High quantities of pDNA were isolated from bacterial cultures using different kits.

3.3.4 Polymerase chain reaction (PCR)
PCR is a simple technique to exponentially amplify defined DNA sequences *in vitro* (Saiki et al., 1988) by means of specific primers. PCR was used for generation and cloning of DNA fragments or for mutagenesis approaches. PCR was carried out in a volume of 50 µl and cycled in a thermocycler (Gen Amp™ PCR System from Applied Biosystems or from Bio-Rad) with the reaction components in the PCR setup listed below.

Methods

10 × Taq buffer	5 µl
Oligonucleotide 1 (10 µM)	1.5 µl
Oligonucleotide 2 (10 µM)	1.5 µl
dNTPs (2 mM)	2- 5 µl
Template DNA	0.1- 5 µg
Polymerase (5 U/ µl)	0.5- 1 µl
H_2O	up to 50 µl

95 °C for 2 min

30 - 35 cycles of
95 °C for 30 seconds
annealing temperature [5] for 30 seconds
68 °C or 72 °C for variable time[6]

Followed by one cycle of
68 °C or 72 °C for 3 to 5 min
Hold at 4 °C - 15 °C

Final elongation was done to fill up 3´- ends of uncompleted PCR products. Five microlitres of PCR setup were analysed by agarose gel electrophoresis for fragment size and approximate DNA quantification.

Site-directed mutagenesis PCR of PhiC31 integrase binding domain

For overlapping PCR reactions, the proofread polymerases KOD or Pfx were used. Reaction conditions were similar to those described above with slight changes. The elongation step was performed at 68 °C and 1- 2 µl $MgSO_4$ was added to the reaction mix. Up to 5 µl Enhancer solution were added to the PCR setup to enhance the amplification and output of PCR. One hundred ng PCR templates (small and large fragments) were used for overlapping PCR. Optimised PCR temperature profiles for first and second PCR to generate small and large products are presented in Table 3.4 and Table 3.5 below.

---

[5] The temperature for primer hybridisation was calculated with the following formula: T_m = no of base pairs (bp) (A+T) × 2 + no of bp (C+G) × 3. The annealing temperature of each primer for optimal hybridization was set as T_m plus 5°C and should be similar between both primers.

[6] Elongation time (polymerisation) depends on the size of the PCR product. Polymerisation of 1 kb product requires 1 min using Taq polymerase.

Table 3.4. Temperature profiles of the first PCR.
Conditions were optimised for different PCR products according to T_M and PCR product size. PCR products were obtained in different sizes according to the mutated amino acid position.

PCR steps	PCR conditions for different fragment sizes		
	762- 640 bp	572- 425 bp	370- 173 bp
initial denaturation	95 °C for 1.50 min	95 °C for 1.50 min	95 °C for 1:50 min
denaturation	95 °C for 0.5 min	95 °C for 0.5 min	95 °C for 0.5 min
annealing	52 °C for 0.50 min	52 °C for 0.75 min	52 °C for 0.67 min
elongation	71 °C for 1.00 min	71 °C for 0:.75 min	71 °C for 0.5 min
final elongation	71 °C for 3.00 min	71 °C for 3.00 min	71 °C for 3.00 min

A second overlapping PCR was performed to anneal the products from PCR 1 and PCR 2, respectively using the outer oligonucleotides to obtain full size construct.

Table 3.5. Temperature profiles for overlapping PCR using proofreading Pfx and KOD polymerases. The PCR program was run for 35 cycles.

PCR steps	PCR conditions for Pfx polymerase	PCR conditions for KOD polymerase
initial denaturation	95 °C for 2.00 min	94 °C for 2.00 min
denaturation	95 °C for 0.33 min	94 °C for 0.33 min
annealing	55 °C for 1.00 min	60 °C for 0.5 min
elongation	68 °C for 1.00 min	72 °C for 0.67 min
final elongation	68 °C for 3.00 min	72 °C for 3.00 min

The PCR product was loaded on an agarose gel and purified upon separation. Purified PCR products were adenylated by Taq polymerase using five µl 10 mM dATP (as the nucleotide donor) and 1 µl T4 PNK in a volume of 50 µl at 37 °C for 30 min. This was done to add adenine overhangs rendering the PCR products compatible to subcloning vectors. After PCI purification adenylated PCR products were immediately subcloned into the pTOPO intermediate vector. Alternatively, the PCR product was separated by gel electrophoresis, extracted, digested by restriction endonucleases *BamHI* and *BstEII* and phosphorylated for subsequent ligation into the final expression vector pCS Int.

The DNA binding domain of PhiC31 integrase (Int) was used as a template for site-directed mutagenesis, which was performed by overlapping PCR. Two PCR products containing the desired point mutation were generated in a first round of PCR (Figure 3.5). Gel-purified PCR products serve as templates in a second round of overlapping PCR with flanking primers. The product of the second PCR comprised the full size binding domain construct, which was digested and inserted into the vector pCS-*NotI*.

Methods

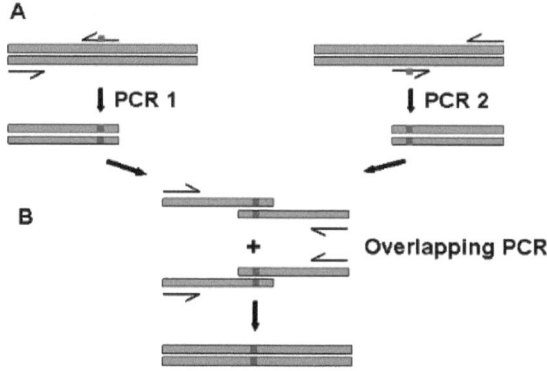

Figure 3.5. Overview of two-step overlapping PCR to generate point mutations.
(A) Fragments were generated in the first PCR using one outer primer and a primer containing a mismatch depicted as a small square. (B) The small and large PCR products include the nucleotide mismatch with the designed point mutation. They were used as templates in an overlapping PCR to generate the full length gene construct.

3.3.5 Design of a linker sequence and construction of a cloning vector pCS+NotI

Two oligonucleotides with *BamHI* and *BstEII* recognition sites at their 5´- ends flanking a *NotI* restriction endonuclease cutting site were annealed by dilution of ten µl 100 mM oligonucleotides with H_2O in a total volume of 100 µl. Oligonucleotides were rapidly boiled and cooled to room temperature. Thirty microlitres of annealed oligonucleotides were phosphorylated in 10 × ligase buffer with 1 µl dATP and 3 µl PNK (phosphate nucleotide kinase) in a 50 µl reaction at 37 °C for 1 hour, PCI-purified, and eluted in 40 µl elution buffer. Linearised dephosphorylated original vector and annealed phosphorylated oligonucleotides were ligated and transformed by electroporation. Positive clones of the backbone vector pCS+*NotI* without the binding domain were verified by restriction digestion using *NotI* and *PstI*, as documented in Section 4.1.1.

3.3.6 Restriction digestion of pDNA and gel electrophoresis

Purified pDNA or purified DNA fragments were digested with restriction endonucleases. Diagnostic restriction enzyme digests were performed in a total volume of 20 µl with 5- 8 µl digested plasmid DNA, 2 µl of the 10 × buffer, 2 µl of 10 × BSA if recommended, 0.5 µl restriction endonucleases and the appropriate volume of sterile H_2O. Restriction digests were performed for 2 hours - overnight at the optimal reaction temperature of the endonuclease. When pDNA digestion required two restriction endonucleases at the same time, whose restriction buffers and recommended cleaving temperatures were different, double digests were first subject to endonuclease cleavage by one enzyme. The digestion mix was purified with the High-pure purification Kit from Roche. Subsequently purified pDNA was digested with the second enzyme.

3.3.7 Isolation of DNA fragments from agarose gels

DNA fragments were separated by gel electrophoresis on a 1 % agarose gel with EtBr (0.5 µg/ml) in 1 × TAE buffer. The electrophoresis was performed at constant voltage of 120 V for 20- 30 min until the migrating DNA samples were sufficiently separated. DNA was visualised by UV light exposure on a transilluminator. The appropriate DNA fragments were excised from the agarose gel and extracted using a Qiagen gel extraction kit according to the manufacturer's instructions. DNA was eluted in 20 µl elution buffer or sterile H_2O.

3.3.8 Dephosphorylation of DNA fragments

Dephosphorylation at 5´- ends of the digested pDNA was performed to avoid self-ligation. Plasmid DNA was dephosphorylated using 0.5 U calf intestinal phosphatase (CIP) at 37 °C for 1 hour. CIP treated vectors were purified from buffers and enzymes with QIAquick Gel extraction Kit (Qiagen) or with the High Pure PCR purification Kit (Roche) according to the manufacturer's instructions.

3.3.9 Ligation of DNA fragments and vectors

DNA ligations were performed with T4 DNA ligase in a total reaction volume of 10-20 µl. Depending on the length of the vector and the insert, the molar ratio of plasmids to insert ratio varied between 1:4 and 1:16. Ligation reactions consisted of 10 × ligase buffer, 1 µl T4 DNA ligase, DNA insert and vector backbone. Reactions were brought to the final volume with sterile H_2O. Ligations were incubated at 16 °C overnight. One to 2 µl of the ligation reaction mix were used to transform *E.coli* by electroporation.

3.3.10 Determination of DNA concentration

The concentration of DNA was determined using Ultrospec 3000 spectrophotometer (Pharmacia Biotech). To determine the concentration of pDNA or genomic DNA, fiftyfold dilution with sterile H_2O was done. The concentration of DNA was determined by the following equation:

$$DNA\ concentration = OD_{260} \times 1000 \times 50\ [ng/ml]$$

The number 50 is the dilution factor, and 1000 is the conversion factor. The ratio extinction of 260 nm to 280 nm (A_{260}/A_{280}) was also measured to determine the purity of DNA. Pure DNA results in extinction ratios of greater than or equal to 1.8.

Methods
3.3.11 DNA sequencing and DNA alignments
To verify DNA sequences 15 µl of DNA at a concentration of 100 ng/µl were sent for DNA sequence analysis to Eurofins MWG Operon Company. Nucleic acid sequence alignments were performed using Clustal W2 software provided by EMBL NBI[7]. Individual sequences were analysed and compared against the database by using the basic local alignment search tool (BLAST), which was provided by National Centre for Biotechnology Information.[8]

3.3.12 Analysis of RNA, isolation and reverse transcription into cDNA
For isolation of total RNA from eukaryotic (293-derived) cells, the RNeasy kit (Qiagen) was used according to the manufacturer's instructions. The concentration of isolated RNA was determined by a spectrophotometer. All steps were carried out on ice and in a cooled centrifuge to avoid RNA degradation. RNA was stored at -80 °C. Transcription of RNA into complementaty DNA (cDNA) was done according to the first strand synthesis protocol from the Protoscript® First Strand cDNA Synthesis kit (NEB) with two µg of purified total RNA.

3.3.13 Quantitative real-time PCR to determine relative DAXX knock down
Gene expression of individual genes can be influenced by RNA interference, which consequently leads to gene-specific down regulation. In principle, specifically designed siRNA binds to a particular ORF in question and consequently inhibits gene expression. The relative down regulation of the target gene can be determined by real-time PCR using reverse transcribed cDNA as a template and appropriate controls. For determination of the standard curve in real-time PCR, a dilution series including five dilutions in duplicates was used. Furthermore, known copy numbers of a control plasmid pDrive+DAXX as an internal standard were used. The construction of the internal control plasmid is described below. The slope of the standard curve was used to calculate the PCR efficiency (1) using either DAXX specific siRNA (target) or GAPDH specific siRNA (reference). The PCR efficiency E was calculated by the ratio R between target and reference (2). The relative down regulation (3) in percentage was derived from the ratios R calculated in (2).

(1) $PCR\ efficiency\ E = 10^{(-1/slope)}$

(2) $rarioR\ (target/reference) = E(target)^{-cp(target)} \times E(reference)^{cp(reference)}$
E represents the calculated PCR efficiency in (1) and cp represents the averaged crossing point values in triplicate obtained by RT PCR.

(3) $relative\ down\ regulation = R(target)/R(reference) \times 100$

---
[7] whttp://www.ebi.ac.uk/Tools/clustalw2/
[8] http://blast.ncbi.nlm.nih.gov/Blast.cgi

Target represents the samples treated with DAXX specific siRNA and reference represents the samples treated with unspecific siRNA (GAPDH specific)

The fluorescence marker Light Cycler Fast Start DNA Master^{PLUS} SYBR Green I (Roche) functions as a DNA-binding dye and can be detected upon intercalation with the amplified DNA (PCR products). The following PCR program for the light cycler was used: pre-incubation at 95 °C for 10 minutes, amplification in 45 cycles at 95 °C for 10 sec, 52 °C for 5 sec and 72 °C for 8 sec.

Construction of control plasmid pDrive+DAXX as an internal standard for qRT-PCR

The genomic sequence from DAXX was derived from the NCBI database. PCR was carried out with Taq polymerase using 200 ng/µl purified genomic DNA derived from 293 cells as template and DAXX specific forward (forw.DAXXqRT) and reverse (rev.DAXXqRT) primers. PCR products in size of 204 bp were inserted into the pDrive Cloning Vector as described in the Qiagen® PCR Cloning Handbook. Insert size was verified by restriction digestion.

3.4 Fluorescence activated cell sorting (FACS)

Flow cytometry describes the emission of optical signals sent by the cells, when passing a laser stream, which is detected by several photomultiplier. The forward scatter detector (FSC) is in line with the light beam and the side scatter detector (SSC) lies perpendicular to it. FSC correlates with the size and SSC depends on the cell's complexity (granularity). The fluorescent detector distinguishes between green fluorescent protein (GFP) expressing cells, cells expressing other fluorescence markers (e.g. mOrange) and non-fluorescent cells. The fluorescence labels (FL), depending on the laser excising the fluorochromes, were FITC for green colour (FL1) and PE for orange colour (FL2). FACS was used to sort cells according to their fluorescent properties upon transfection of plasmids carrying genes expressing fluorescent proteins.

Two different transfection setups using different plasmids were carried out with subsequent FACS analysis:

a) Establishment and evaluation of stable GFP-containing reporter cell lines

293 cells were stably transfected with the plasmid p*attP*-polyA-*attB*-EGFP (Figure 4.16) in a 6-cm dish with a ratio FuGENE [µl]: DNA [µg] = 2:1. Two weeks after selection in cell culture D-MEM medium supplemented with 500 ng/ml neomycin, clones were transferred into plates with fresh medium and expanded to an appropriate amount of cells to establish

Methods

293-based GFP containing reporter cell lines. Established GFP-containing reporter cell lines were transfected with five µg pDNA encoding either an active (wt Int) or inactive (mInt) version of integrase to analyse fluorescence activity of GFP. GFP is activated upon integrase-mediated out-recombination of the terminating polyA signal.

b) Evaluation of integrase mutants for recombination activity in a GFP-containing reporter cell line obtained from transfection setup a

The GFP-containing cell line (established as described above) was co-transfected with integrase-encoding plasmids and mOrange expressing plasmid in one well of a 6-well plate using a weight ratio of 2:1, which corresponds to 1333 ng integrase plasmid and 666 ng mOrange plasmid.

Forty-eight hours post-transfection cells were harvested by washing once with D-PBS, centrifuged at 2000 rpm for 5 min at 4 °C and subsequently resuspended in appropriate amount of FACS buffer (D-PBS with 1 % FBS), transferred into an Eppendorf tube and stored on ice. Up to 10000 cell counts (equals to 100 %) per sample were analysed by flow cytometry using a FACS Canto. The fractions of each sample containing non-fluorescent cells, eGFP expressing cells, or eGFP and mOrange expressing cells were quantified. Transfection efficiencies were only corrected in the second transfection setup by co-transfection with mOrange, assuming an equal plasmid uptake per cell. The relative GFP expression level was recorded as a measure for excision activity within a genomic context.

3.5 Dual luciferase assay

The dual luciferase assay (Promega) was performed to measure the excision activity of wt Int and its mutants. HEK 293 cells were grown on 24-well plates and were at the same time co-transfected with three different plasmids 48 hours before luciferase measurement. The transfection setups are described in detail in Table 3.6. The first plasmid pCS Int encodes the integrase enzyme or its mutant derivative. The second plasmid pLucCR contains the CMV promoter-driven firefly luciferase reporter gene downstream of the Int-specific excision construct, in which a termination site (polyA sequence) is flanked by the *attB* and *attP* recognition sites. The third plasmid pRL-TK encoding a tymidine kinase promoter driven *Renilla* luciferase gene is co-transfected as an internal control for transfection efficiency. Alternatively, a plasmid was co-transfected encoding human codon-optimised *Renilla* luciferase (phRL-O).

Methods

Table 3.6. Experimental outline of transfection conditions of luciferase assays performed. The initial experiment was performed with the conditions stated in Setup 1 to analyse the effect of the integrase plasmid concentration on DNA excision.

Parameters	Setup 1	Setup 2	Setup 3
plasmid ratio pLucCR : Int	1 : 0.9	1 : 1	1 : 5
total DNA transfected in ng per triplicate in 24-well scale	1200	2040	2040
integrase plasmid (pCS Int)	600	400	1600
pLucCR (+hs (hotspot)) in ng	540	400	400
pRL-TK in ng	60	-	-
phRL-O in ng	-	40	40
stuffer plasmid pUC19 in ng	-	1200	-
amount of transfection reagent: DNA complex per well in µl	20	33	33
amount of transfected cells used for luciferase assay in µl	5	10	10
with injector, Δt = constant	no	yes	Yes
manually	yes	no	No

One hundred microlitres of freshly prepared 1 times passive lysis buffer were added to each well of a 24-well plate. The cells were incubated at room temperature for 10 min. Ten microlitres of each cell lysate were directly pipetted into a 96-well plate (Nunc). The remaining cell lysates were transferred into Eppendorf tubes and rapidly frozen in liquid nitrogen.

Thirty-five microlitres LAR II were injected automatically per well to initiate the firefly luciferase reaction. After 10 seconds, 35 µl Stop&Glo reagent were added automatically by injectors. Simultaneously, with quenching firefly luciferase activity, R*enilla* luciferase was activated. Luciferase activities were measured in a plate-reading Microlumat Plus LB 96V Luminometer.

3.6 Animal studies

3.6.1 Hydrodynamic tail vein injection

Female C57BL/6 mice derived from a common inbred strain of lab mice were housed in the animal facility and treated according to the regulations of the Government of Upper Bavaria. Mice were injected by hydrodynamic plasmid delivery via the tail vein over 6 seconds using a 3-ml syringe with a 28-G needle. Twenty micrograms of Int encoding plasmids and 20 µg substrate plasmids containing the *attB* site and the expression cassette for human blood coagulation Factor IX (hFIX) were injected. Total volume was

Methods

1.8 ml of 0.9 % NaCl. Cell cycles of murine liver were induced by intraperitoneal administration of 50 µl carbon tetrachloride (CCl_4) diluted in mineral oil at a ratio of 1:1.

3.6.2 Measurement of alanine aminotransferase (ALT)

ALT is normally present in high concentrations in the liver. Elevated ALT levels in the blood usually indicate a liver disease or a hepatic injury, which should be observed after tail vein injection of foreign DNA. For quantitative determination of ALT in murine serum as a signal for liver toxicity, the ALT kit from Randox was used according to the manufacturer's instructions. Blood samples were taken retro-orbitally from mice, collected in Eppendorf tubes and centrifuged for 2 min at 10000 rpm between 5 °C and 10 °C. Supernatant containing mouse serum was collected and stored at 4 °C until ALT assay was carried out by applying 15 µl serum per mouse with 90 µl substrate buffer/enzyme mix. ALT activity corresponds to absorbance change over time detected at 340 nm in a photo spectrometer.

3.6.3 Enzyme Linked Immunoabsorbent Assay (ELISA)

Detection of hFIX concentration in murine serum was performed using a sandwich ELISA. For the generation of the standard curve, purified hFIX (ProSpec-Tany Technogene) was used in final concentrations between 3200 ng/ml and 3 ng/ml. 96-well plates (Nunc Maxi Scorp) were coated with an anti-hFIX antibody (Sigma). The anti-hFIX antibody was used in a dilution of 1:10000 in coating buffer (0.1 M $NaHCO_3$, pH 9.4) and plates were incubated with 50 µl per well at 4 °C overnight. Plates were washed twice with 200 µl dilution buffer TBS-T per well. After removing TBS-T, 200 µl dilution buffer (TBS-T+ 5 % BSA) were used to block the wells for one hour at room temperature. Plates were washed twice with TBS-T. Murine serum samples were diluted in appropriate relations with dilution buffer. After blocking, plates were loaded with 50 µl per well and incubated at 37 °C for two hours. Plates were washed twice with TBS-T. For detection, an anti-HRPO (horseradish peroxidase) antibody (Biozol) was used. Fifty microlitres of an HRPO conjugated secondary antibody were added to each well as a 1:1000 dilution. Plates were incubated for one hour at 37 °C. Plates were washed three times with TBS-T. Fifty microlitres substrate solution (OPD Sigma Fast) were added for final development incubating at room temperature for about 10 min. The reaction was stopped and enhanced by adding 50 µl 1 M H_2SO_4. Samples were read at 492 nm in a plate reader (Tecan, Magellan3 software).

4. Results

4.1 Construction of integrase mutants by site-directed mutagenesis

The aim of this work was to increase the integration efficiency of PhiC31 integrase by site-directed mutagenesis of the PhiC31 integrase binding domain (BD). Mutations were introduced by replacing charged amino acids with alanine, an approach named alanine scanning. The PhiC31 BD has not yet been screened in a large-scale mutagenesis study using alanine scanning. Since beneficial mutations were found within the same domain of the Sleeping Beauty transposase (Yant et al., 2004), alanine scanning was chosen for the BD of PhiC31. The integrase BD was analysed for charged amino acids, such as aspartic acid (D), glutamic acid (E), lysine (K) and arginine (R). From the 41 charged amino acids identified, 22 amino acid positions (53 %) were mutated to alanine (A). The outline of the cloning strategy for the generation of plasmids with individual point mutations in the DNA sequence of the PhiC31 integrase DNA-binding domain is presented in Figure 4.1.

Preparation of truncated cloning vector pCS+*NotI* without binding domain (4.1.1)	Generation of mutated PCR product (insert) by overlapping PCR method (4.1.2)
Digestion of pCS-Int with *BamHI* and *BstEII* to remove binding domain	First round of PCR to construct small and large mutated fragments
Design of a short linker sequence containing *NotI* and compatible flanking restriction sites	Second round of PCR using overlapping PCR of small and large mutated fragments to create binding domain with desired point mutation
Insertion of linker sequence into linearised backbone vector pCS by ligation	Subcloning of full length mutated binding domain into pTOPO vector
Linearisation of truncated vector pCS+*NotI* with *BamHI* and *BstEII*	Isolation of mutated binding domain from intermediate vector pTOPO with *BamHI* and *BstEII*

Generation of plasmids encoding the mutated integrase gene (4.1.3)
Ligation of open pCS vector and insert containing mutated binding domain and transformation, plasmid isolation and verification of mutation by sequencing

Figure 4.1. Cloning strategy for the construction of integrase point mutants.
The outline describes the preparation and construction of the vector (Section 4.1.1) and the insert (4.1.2) with single point mutations within the PhiC31 integrase binding domain and the subsequent generation of plasmids encoding the mutated integrase gene (4.1.3).

Results

4.1.1 Preparation of truncated cloning vector pCS+NotI

The original expression vector pCS Int encoding the full-length wt integrase gene was digested with restriction endonucleases *BamHI* and *BstEII* to remove the DNA binding domain. With insertion of a short linker sequence into the linearised vector at the respective cutting sites the truncated vector pCS+*NotI* was constructed for simplified cloning to avoid false positive clones encoding wt integrase. The design of the linker sequence and the construction of the cloning vector pCS+*NotI* were described in Section 3.3.5. Agarose gel electrophoresis documented the restriction digestion of the truncated vector pCS+*NotI* (Figure 4.2.A). Vector maps of pCS Int and pCS+*NotI* supporting the cloning strategy are illustrated in Figure 4.2.B and 4.2.C, respectively.

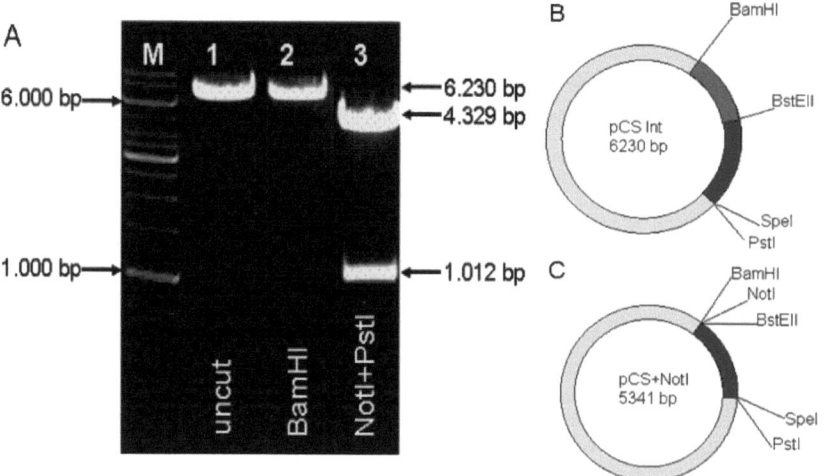

Figure 4.2. Agarose gel electrophoresis of the vector backbone pCS+*NotI* uncut, *BamHI* digested, and *NotI*+*PstI* digested and respective vector maps.
The original expression vector pCS Int was modified by replacing the sequence encoding the native binding domain (BD) between *BamHI* and *BstEII* with a short oligonucleotide sequence containing a *NotI* site to construct the pCS+*NotI* backbone vector for simplified cloning. (A) Agarose gel showing restriction digest of respective plasmids. The restriction pattern in line 1 and 2 show the pCS Int vector uncut and *BamHI* restricted in size of 6230 base pairs (bp), the restriction pattern in line 3 shows the digest of the backbone vector in size of 5341 bp with *NotI* and *PstI* resulting in two fragments in size of 4329 bp and 1012 bp, respectively. M= 1 kb DNA marker. (B) Vector map of pCS Int. *BamHI* and *BstEII* are flanking the BD, *BamHI* and *SpeI* flank the entire integrase coding sequence. (C) Vector map of pCS+*NotI*. The *NotI* site is inserted instead of the BD.

Results

4.1.2 Generation of mutated integrase binding domain by overlapping PCR

In the first round of PCR, oligonucleotides containing either a *BamHI* or *BstEII* site together with one mismatch oligonucleotide were used to yield mutated PCR products by overlapping PCR (Figure 3.2.A). The restriction endonucleases *BstEII* and *BamHI* are unique cutters in the pCS Int vector flanking the DNA binding domain within a region of 910 bp. PCR products containing 6 point mutations of PCR 1 and PCR 2 are presented in Figure 4.3.

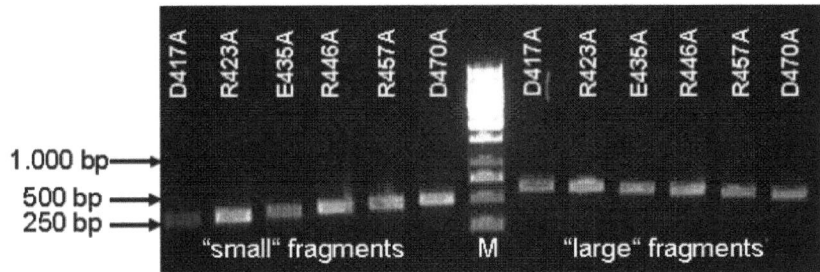

Figure 4.3. PCR products obtained after the first round of PCR with mutagenic primers were analysed by agarose gel electrophoresis.
PCR products were loaded on a gel together with the 1 kb DNA marker (M). Three highlighted bands in size of 1000, 500, and 250 bp within the marker help to identify the fragment size of the samples. The PCR products contain individual point mutations as indicated in the figure. The small and the large fragments were used as template in overlapping PCR.

The PCR products (Figure 4.3) were used for overlapping PCR resulting in full-length PCR products with the mutated DNA binding domain sequence (Figure 4.4).

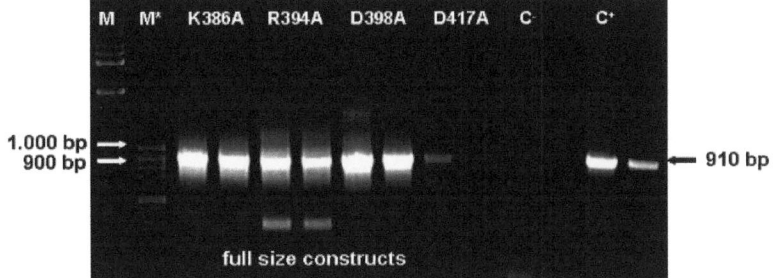

Figure 4.4. Analysis of second round PCR products by agarose gel electrophoresis.
PCR products including the mutated binding domain (BD) are shown for four mutants. Controls were included: C⁻ contains no template, C⁺ represents amplification of pCS Int with outer primers resulting in a band of 910 bp. M = 1 kb DNA marker, M* = 100 bp DNA marker. Two highlighted bands in size of 1000 and 900 bp within the marker help to identify the size of the samples. PCR products comprising the full size construct in size of 910 bp were subcloned into intermediate pTOPO vector.

Results
4.1.3 Generation of plasmids encoding the mutated integrase gene

The insert containing the mutated integrase binding domain was isolated from the pTOPO vector and inserted into *BamHI* and *BstEII* linearised pCS+*NotI* vector as described in Figure 4.1. The final set of 22 plasmids contained point mutations within the integrase binding domain and were named based on the mutation positions (Table 4.1).

Nucleotide and amino acid sequences with the native integrase BD are shown in Figure 4.5. The amino acid positions which were changed to alanine are represented in bold and underlined text. Integrase BD mutant vectors were confirmed by DNA sequence analysis.

```
atggacaagctgtactgcgagtgtggcgccgtcatgacttcgaagcgcggggaagaatcg
 M  D  K  L  Y  C  E  C  G  A  V  M  T  S  K  R  G  E  E  S
atcaaggactcttaccgctgccgtcgccggaaggtggtcgacccgtccgcacctgggcag
 I  K  D  S  Y  R  C  R  R  R  K  V  V  D  P  S  A  P  G  Q
cacgaaggcacgtgcaacgtcagcatggcggcactcgacaagttcgttgcggaacgcatc
 H  E  G  T  C  N  V  S  M  A  A  L  D  K  F  V  A  E  R  I
ttcaacaagatcaggcacgccgaaggcgacgaagagacgttggcgcttctgtgggaagcc
 F  N  K  I  R  H  A  E  G  D  E  E  T  L  A  L  L  W  E  A
gcccgacgcttcggcaagctcactgaggcgcctgagaagagcggcgaacgggcgaacctt
 A  R  R  F  G  K  L  T  E  A  P  E  K  S  G  E  R  A  N  L
gttgcggagcgcgccgacgcccttgaacgcccttgaagagctgtacgaa
 V  A  E  R  A  D  A  L  N  A  L  E  E  L  Y  E
```

Figure 4.5. Nucleotide sequence and amino acid sequence of the integrase binding domain. Underlined and bold letters represent the 22 residues changed to alanine to construct the respective mutants. The first letter M represents amino acid position 365. The last letter E represents the amino acid position 480 within the amino acid sequence of the integrase. The nucleotide sequence was translated using the translate tool at www.expasy.ch/tools/dna.html.

Exclusively charged amino acids were selected to be mutated towards the uncharged and small amino acid alanine to generate alanine substitution mutants. The alanine scanning approach represents a convenient method in enzyme mutagenesis. The domain to be mutagenised within the integrase ORF was the DNA binding domain since this domain has not yet been screened for efficiency mutations in a large-scale mutagenesis using alanine scanning and beneficial mutations were found within the same domain of the Sleeping Beauty transposase (Yant et al., 2004). The peptide sequence of the integrase BD was screened for charged amino acids, such as aspartic acid (D), glutamic acid (E), lysine (K) and arginine (R). From the 41 charged amino acids, 22 amino acid positions (53 %) were mutated towards alanine (A) by altering the genetic code. Mutations of the integrase BD were confirmed by DNA sequence analysis. All constructed integrase single mutants presented in Table 4.1 were screened for altered recombination proficiency. At first, integration efficiency of Int mutants was investigated in HeLa cells.

Results

Table 4.1 Constructed PhiC31 integrase mutants.
The first letter indicates the amino acid changed to alanine (A). Aspartic acid represents the letter D, glutamic acid E, lysine K and arginine R. The number within the ORF indicates the amino acid position which was changed.

PhiC31 integrase point mutants*	
D366A	E406A
K367A	D417A
E371A	R423A
R380A	R429A
E382A	E432A
E383A	E435A
K386A	R446A
R390A	K450A
R393A	K457A
R394A	R461A
D398A	D470A

*__General comment:__ the 22 designed mutants within this thesis represent a different amino acid position (**4 amino acid shift**) than the PhiC31 integrase mutants within my publication (Liesner et al., Human Gene Ther., 2010). The mutant D366A is referred to D470A, K367A is referred to K371A, D470 is referred 474A and so on.

4.2 Integration efficiency in HeLa cells of PhiC31 integrase mutant derivatives

The rationale of this experiment was to determine the integration efficiency of the 22 PhiC31 integrase mutants in mammalian cells. The integration efficiency is directly correlated with the number of clones surviving selection after transfection. Experiments were carried out by co-transfecting HeLa cells with plasmids coding either for the wt Int, non-functional Int (mInt) or for each of the newly constructed Int mutants and the donor plasmid p7. The number of colonies, being used as a measure for integration, was determined by a method termed colony-forming assay (CFA), schematically illustrated in Figure 4.6. The used molar ratio of Int:p7 was 0.5:1 using 2000 ng pDNA. No additional stuffer DNA was included.

Results

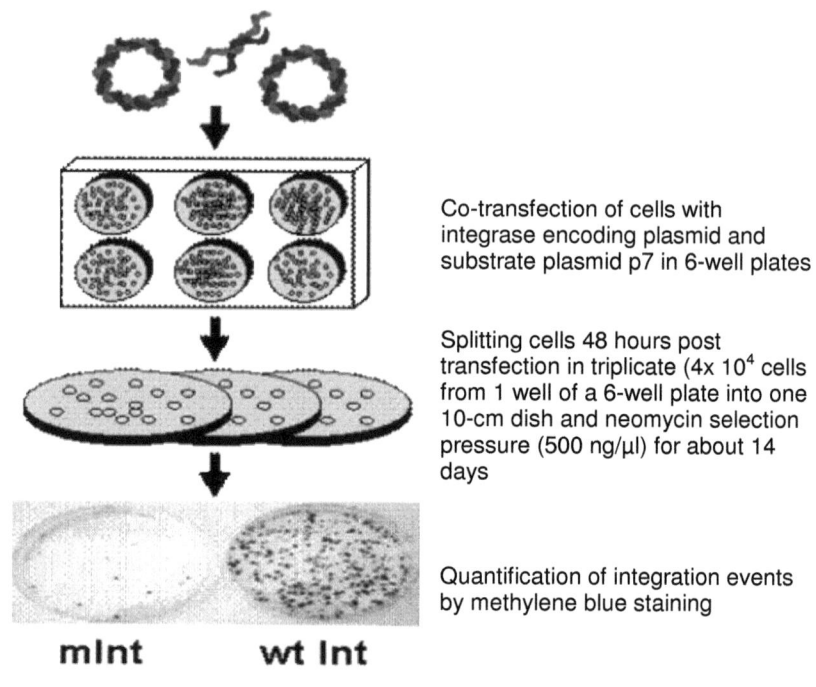

Figure 4.6. Schematic overview over a colony-forming assay (CFA).
The text to the right describes the procedure of the assay. mInt designates the cells transfected with non-functional Int (negative control). wt Int designates the cells transfected with wild type PhiC31 Integrase encoding plasmids.

Colony numbers obtained after wt Int mediated integration were compared to colony numbers obtained by individual Int mutants as a measure of the relative integration efficiency. The integration efficiency was calculated from the absolute number of integration events by setting the colony number obtained after wt Int-mediated integration to 100 %. The integration efficiencies of the 22 different single amino acid substitution mutants are presented in Figure 4.7.

Five integrase mutants, D366A, E382A, E383A, K457A and D470A, showed an integration efficiency between 1.2 and 1.7-fold higher than wt Int. Seven mutants (K367A, R380A, R393A, R394A, D417A, E432A, and K450A) lost their recombination activity into *pseudo attP* (*attP`*) sites. The integration efficiencies of these mutants were as low as the negative control mInt, which was about seven times lower than wt Int mediated integration. Ten Int mutants, E371A, K386A, R390A, D398A, E406A, R423A, R429A, E435A, R446A, and R461A, showed integration efficiencies comparable to wt Int-mediated integration.

Results

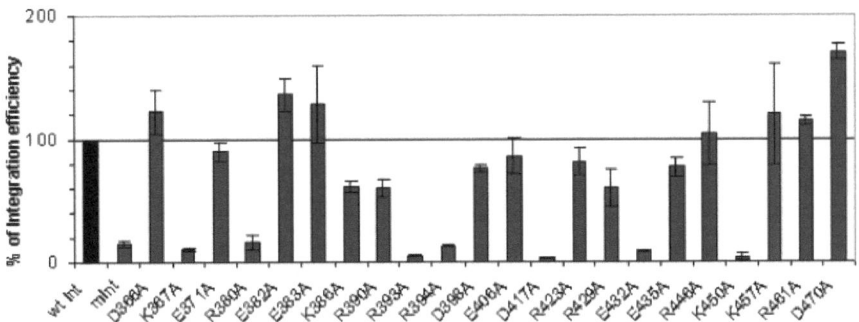

Figure 4.7. Integration efficiency of all constructed integrase mutants obtained by colony-forming assay (CFA) in HeLa cells.
The molar ratio of p7: Int = 1:0.48, which equals to a plasmid amount of 1000 ng p7 and 1000 ng integrase encoding plasmid was used (Table 3.2). Effect of amino acid substitutions in the integrase binding domain on the integration efficiency is shown. HeLa cells were co-transfected with plasmids encoding the wt integrase (wt Int), the non-functional version (mInt) or with integrase missense mutant together with a substrate plasmid p7 encoding neomycin and the *attB* site. Mean values are shown with error bars indicating deviation within a triplicate of one mutants.

In the first recombination screening in HeLa cells, only mutant D470A had an increase in integration efficiency of 1.7. In the next step towards improving integration efficiency, the plasmid ratio of integrase encoding plasmids was addressed in dose dependent studies.

4.3 Dose dependent studies with different amounts of PhiC31 integrase plasmids

The following studies were performed to evaluate to what extent the ratio between the Int expressing plasmids and the substrate plasmid p7 would influence the formation of colonies. Therefore, I wanted to analyse, whether and how the overall integration efficiency is influenced by the dose of Int within a cell. This was done by varying the plasmid ratios as described in Table 3.2.

Four mutants, D366A, E383A, K457A and D470A, showing improved integration efficiency within the first assay at a molar ratio of p7:Int = 1:0.48 were used for dose-dependent studies in HeLa cells (Figure 4.8.A) to evaluate the effect on colony formation at different plasmid ratios. HeLa cells were co-transfected with only 50 ng substrate plasmid p7 and 102 ng Int plasmid (molar ratio 1:1) or with 513 ng Int plasmid (molar ratio 1:5) and with additional amount of stuffer plasmid pBS, respectively (Table 3.2). The cells transfected with high Int plasmid dose p7:Int = 1:5 resulted in 1.3-fold increase in integration events compared to low amounts of Int transfected (1:1 ratio).

Results

The initial integration efficiency of mutant D366A in the overall CFA was 122 % at a 1:1 ratio without stuffer plasmid. With the new conditions, the mutant had a 68 % integration efficiency at a 1:1 ratio and decreased slightly to 61 % at a 1:5 ratio.

The mutant E383A was slightly better with 107 % integration efficiency and 111 % integration efficiency at 1:1 and 1:5 ratios, respectively; the original result was a 1.28-fold increase (Figure 4.1). The mutant K457A had an initial efficiency of 120 % (Figure 4.8.A), but was 65 % at a 1:1 ratio and increased more than twice to 135 % at a 1:5 ratio.

The mutant D470A, which showed the most promising integration efficiency (1.7-fold) at the initial ratio (Figure 4.5) but had an integration efficiency of 121 % (1:1) and 104 % (1:5) compared to wt. The evaluation of the CFA showed very few differences in integration efficiency between the plasmid ratios in mutants D366A, E383A and D470A.

In another experiment, the plasmid ratios were changed to p7:Int = 1:1 and 1:20 (=1950 ng Int plasmid) (Figure 4.8.B). Wild type integrase mediated integration efficiency was increased 1.3-fold when increasing the Int plasmid 20 times. The mutants K457A and D470A showed about 1.5-fold improvements at equimolar ratios (Figure 4.8.B). At high Int dose (1:20 ratio) the fold increase of K457A and D470A was only slightly higher compared to wt Int, with an integration efficiency of 1.4 times and 1.5 times, respectively, compared to 1.2-fold improvement with the wt Int. In the CFA experiments (Figure 4.8.B) two independent experiments with almost identical outcome were combined to verify the improved integration efficiency.

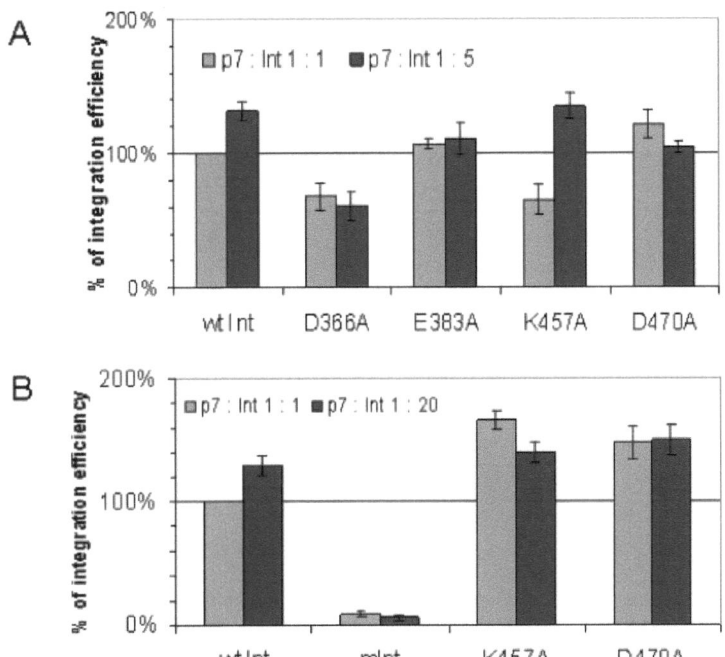

Figure 4.8. Dose dependent studies with increasing plasmid transfection ratios of selected integrase mutants.
Colony-forming assay of the integrase mutants D366A, E383A, K457A, and D470A in comparison to wt integrase and mInt in HeLa cells is shown. For each transfection 50 ng substrate plasmid p7 were used. For the different plasmid ratios, respective plasmid amounts in ng are listed in Table 3.2. (A) For plasmid transfection the molar ratios of p7 to integrase plasmid were 1:1 and 1:5, respectively. (B) The molar ratios of p7: Int plasmid were 1:1 and 1:20, respectively. In the second experiment, transfections were performed twice in triplicate.

Next, the effect of lowering the integrase dose on integration efficiency was analysed. Dose dependent studies were performed in which the substrate plasmid p7 was transfected in surplus compared to the Int plasmid. The plasmid amounts in favour of p7 were twofold, tenfold and fortyfold higher compared to the integrase plasmid (Figure 4.9). Substrate plasmid p7 was transfected in constant amounts (50 ng) and stuffer plasmid was used for compensation for missing plasmid volume. The molar ratios of transfected plasmids with respective amounts in ng are described in Table 3.2.

The integration assay clearly showed that with decreasing amounts of Int plasmid transfected, the number of G418 resistant colonies also decreased. Integration events with a transfection ratio of p7:Int = 1:0.5 were higher for each Int mutant than integration events at ratios p7: Int = 1:0.1 and 1:0.025, respectively. Comparing the absolute number of colonies with the previous assays confirmed the result (data not shown).

Results

Mutant E383A showed a more than twofold increase in colony formation compared to wt integrase at a ratio p7:Int = 1:0.5. The fold decrease from the initial plasmid ratio (1: 0.5) to the lowest transfection ratio (1:0.025) was 20 % for wt Int, about 140 % for mutant E383A, 100 % for mutant K457A and 40 % for mutant D470A. However, the overall number of resistant colonies was very low and ranged between 5-16 colonies per 10-cm dish.

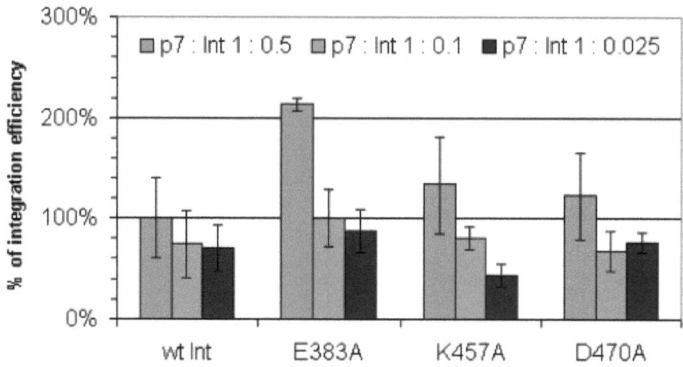

Figure 4.9. Dose dependent studies with decreasing amounts of integrase plasmid compared to substrate plasmid p7.
The integrase mutants E383A, K457A and D470A were compared with wild type integrase. The plasmids were transfected in three different transfection ratios (p7: Int) = 1:0.5, 1:0.1 and 1:0.01 into HeLa cells and integration efficiency was evaluated by CFA. The experiment was carried out in triplicate.

The integration efficiency of single mutants in HeLa cells could not be significantly improved compared to the wt integrase. Therefore, additional strategies to further improve integration efficiency involved the construction and testing of double mutants based on the integration efficiencies obtained so far.

4.4 The effect of double mutants on integration efficiency

After the first analysis using single mutants, additional attempts to increase the integration efficiency further were performed. Therefore, single Int mutants showing increased efficiency in HeLa cells were used to generate double mutants. The strategy was to select the mutant D470A showing the most improved integration efficiency in the first screen in HeLa cells and use it in combination with other single mutants, showing slightly higher integration efficiencies than wt integrase. The double mutants E382A+ D470A and K457A+ D470A were constructed and tested.

Selected mutants and two selected double mutants were tested for integration efficiency in HeLa cells by CFA using low amounts of Int and substrate plasmids. The integration efficiency is presented in Figure 4.10. All tested mutants showed enhanced integration efficiency compared to wt Int. The single mutants D470A and E382A obtained about 1.4 times better integration efficiency compared to wt Int. The double mutant E382A+ D470A showed synergistic effects with a fold improvement of 1.7 times compared to wt Int. Combining the mutation K457A with mutant D470A had a negative effect on integration. The double mutant K457A+ D470A showed only a slight improvement. The negative control group mInt dropped about 6 times compared to the wt integrase.

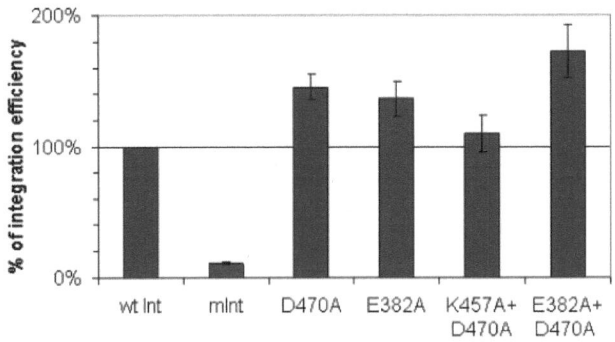

Figure 4.10. Integration efficiencies of single and double mutants in HeLa cells.
The plasmid ratio for transfection was p7: Int = 1: 0.5. The remaining amount of DNA was transfected with stuffer plasmid pBS. Plasmid ratios for transfection are listed in table 3.4. Experiments were carried out in triplicate.

Next, the effect of enhanced Int plasmid concentrations of the two double mutants shown in Figure 4.10 was investigated. Plasmid transfections with two different plasmid ratios were carried out (Figure 4.11) in triplicate.

The first experiment (#1) shows relative integration events after transfection of two different Int plasmid doses, p7:Int = 1:0.5 and 1:20. The mutant K457A+ D470A obtained almost the same integration efficiency as wt Int. The mutant E382A+ D470A obtained 1.7 times higher integration efficiency than wt Int at a low Int plasmid dose with a plasmid ratio p7:Int = 1:0.5 (left bars). Increasing the Int plasmid dose twentyfold leads to a drop to 85 % of wt Int. The double mutants K457A+ D470A (110 % at a 1:1 ratio and 265 % at a 1:20 ratio) and E382A+ D470A (170 % at a 1:1 ratio and 235 % at a 1:20 ratio) however, showed improved integration efficiency at a high Int plasmid dose.

These results were confirmed by repeating the integration efficiency twice in experiments #2 and #3. This resulted in an averaged fold-increase of 3 times and 3.3 times of the mutant K457A+ D470A and E382A+ D470A (black and dashed bars), respectively,

Results

compared to the wt integrase. The integration efficiencies for mutant K457A+ D470A were within 30 % deviation for all three independent experiments (#1, #2, and #3). The integration efficiency of integrase mutant E382A+ D470A shown in experiment #2 was almost twofold higher than obtained from experiment #1 and #3, which is inconsistent. The integration efficiency of the negative control mInt, representing random integration events, included only in experiments #2 and #3 was approximately 5 times lower than wt Int.

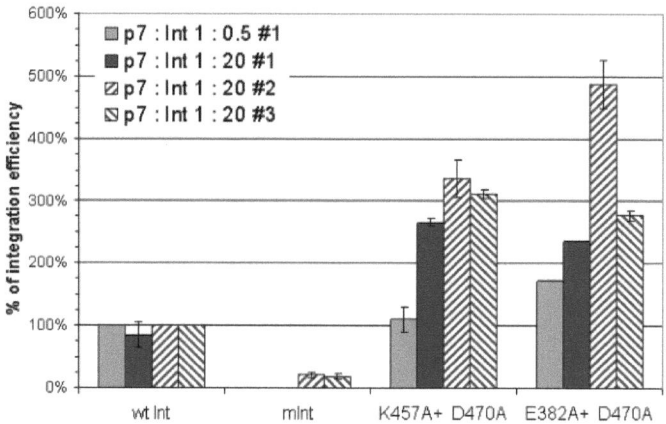

Figure 4.11. Integration efficiencies of double mutants in HeLa cells at two different plasmid ratios p7:Int = 1:0.5 and 1:20.
The CFA was performed to compare double mutants K457A+ D470A and E382A+ D470A against wt integrase at a low and a high dose of transfected integrase plasmid. Experiment #1 shows integration efficiency after high and low integrase dose, experiment #2 and #3 were independently performed to confirm the tendency of increased integration mediated by the double mutants. The three independent experiments were performed in triplicate.

4.5 Integration efficiencies of integrase mutants in cell lines of different origin

Since the single and double mutants demonstrated improved integration efficiencies in HeLa cells, I wanted to evaluate the mutants within the context of other human cell lines. Therefore, the human cell lines HCT, Huh7, 293 and the mouse cell line Hep1A were transfected with mutant plasmid DNA. In the context of these cell lines, individual integrase mutants were evaluated for their integration efficiencies and compared to wt integrase. The efficacy of a mutant in different cell lines was not expected to be comparable, due to cellular cofactors and chromosomal context effects influenced by transcription patterns.

4.5.1 Integration efficiencies in HCT cells
Human colon-derived HCT cells were transfected with selected integrase mutants at high and low plasmid ratios to analyse the effect and influence of integrase mutants and

Results

plasmid dose on the integration efficiency (Figure 4.12). The mutants K457 and D470A showed a 1.5-fold improvement of integration efficiency at a high plasmid ratio compared to wt Int at 1:1 ratio. At the equimolar plasmid ratio (1:1) however, the mutant K457A had almost twofold integration efficiency and the mutant D470A dropped to 42 % of the wt Int. The two double mutants had different effects upon increasing the plasmid ratio twentyfold with integrase encoding plasmid. The double mutant K457A+ D470A showed a decline from 143 % (at a 1:1 ratio) to 115 % (at a 1:20 ratio), while the second double mutant E382A+ D470A showed a slight increase from 127 % (at a 1:1 ratio) to 144 % (at a 1:20 ratio) in integration efficiency.

A second CFA was performed to confirm these results shown in Figure 4.12 (data not shown). This assay showed between 2.5-fold and fourfold higher integration efficiencies of the mutants compared to the wt integrase. The number of colonies, transfected with wt integrase encoding plasmid (control group) was with an average of 33 very low, compared to 144 obtained from the raw data in Figure 4.12.

Due to this large inconsistency between the control groups of both integration assays the second assay could not be properly evaluated, although the number of colonies obtained with several mutant integrases were evaluable and within the expected range of colonies. An additional confirming experiment should be performed.

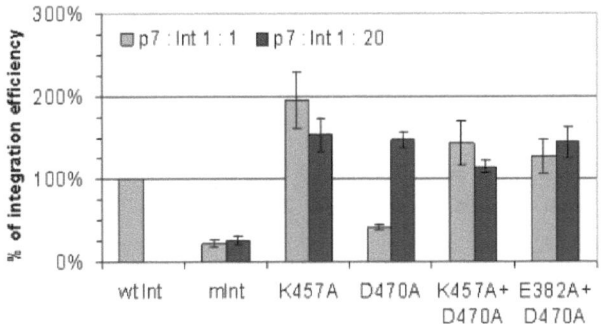

Figure 4.12. Integration efficiencies of selected integrase mutants in HCT cells.
Two integrase single mutants K457A and D470A and two double mutants K457A+ D470A and E382A+ D470A were screened. Co-transfection was performed at a ratio of p7:Int = 1:1 and 1:20. Experiments were carried out in triplicate.

4.5.2 Integration efficiencies in Huh7 cells

Integration assays at high integrase plasmid doses p7: Int = 1:20 in liver-derived Huh7 cells were carried out (Figure 4.13). The mutant K457A analysed twice, showed 70 % integration efficiency compared to wt Int. The mutant D470A had no improvement

Results

compared to wt integrase although the results had a high standard deviation. The double mutants K457A+ D470A and E382A+ D470A displayed 1.5- and 2.3-fold improvements, respectively, compared to the wt integrase. The negative control showed about 20-fold lower integration efficiency.

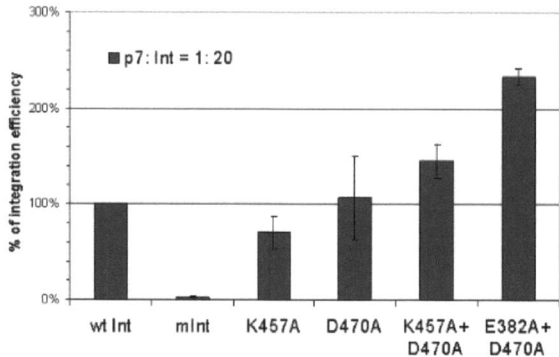

Figure 4.13. Integration efficiencies of selected integrase mutants in Huh7 cells.
For co-transfection a plasmid ratio of p7:Int = 1:20 was used. The mutants K457A, D470A and the double mutants K457A+ D470A and D470A+ E382A were tested. Two independent experiments were performed with wt integrase and the mutants K457A and K457A+ D470A in triplicate. The values were averaged.

4.5.3 Integration efficiencies in HEK 293 cells

Integration efficiencies of several integrase mutants and a mouse codon-optimised derivative of the wt PhiC31 integrase were evaluated in 293 cells using both low and high amounts of Int plasmids at ratios p7:Int = 1:1 and 1:20, respectively (Figure 4.14). The synthesised codon-optimised version of the PhiC31 integrase with a reduced number of CpG dinucleotides to avoid gene silencing showed promising recombination activity in an embryonic stem cell-derived mouse strain (Raymond and Soriano, 2007). The codon-optimised version was only used in 293 cells and was tested in luciferase assays also, performed in 293 cells (data not shown).

To test whether the codon-optimised integrase also shows positive effect on integration efficiency within the context of human-derived 293 cells, this derivative was included in the recent experiment. The integration efficiencies of the mutants K457A, D470A and K457A+ D470A were with 1.17, 1.09 and 1.19-fold rather similar to the wt Int. The double mutant E382A+ D470A showed about 1.5-fold improvement, and the included codon-optimised integrase (Int c.o.) showed only the same integration efficiency as wt integrase. The mutant D470A with high efficiency in HeLa cells showed a very low improvement at an equimolar plasmid ratio (1:1) comparable to wt levels. At a high plasmid dose, 1.5-fold higher integration was observed in D470A compared to wt Int with 1.2-fold (right bars at

1:20 ratio). The improvement in integration efficiency in both wt Int and D470A was 1.24-fold when shifting the plasmid ratio from p7:Int = 1:1 to 1:20. The mInt was about 5 times lower than wt Int (Figure 4.14).

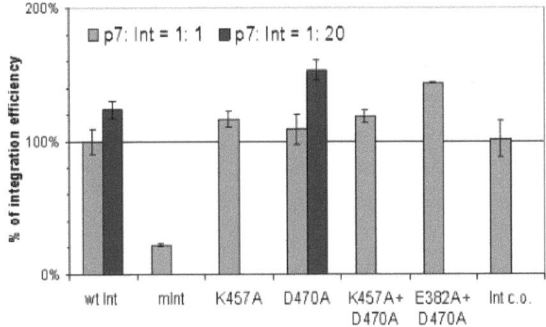

Figure 4.14. Integration efficiencies of selected integrase mutants in 293 cells.
Co-transfection of plasmids was carried out using plasmid ratios p7: Int = 1: 1 for all mutants tested. The plasmid ratio p7:Int = 1:20 was used for wt Int and mutant D470A within the same experiment. At equimolar ratios (1: 1) 85 % stuffer plasmid pBS was used for transfection. The codon-optimised version of integrase (Int c.o.) was included. Two triplicate of each group were used in the experiment.

4.5.4 Integration efficiencies in Hep1A cells

Three mutants have been tested in the mouse derived Hep1A cell line (Figure 4.15). The plasmid ratio used for transfection was p7:Int = 1:10 and 50 ng p7, and 1950 ng pCS Int plasmids were transfected twice in duplicate. No additional stuffer plasmid was used. The mutants K457A and E382A+ D470A showed a 1.3-fold increase in colony formation compared to the wt integrase. The double mutant K457A+ D470A (63 %) showed only a slight improvement compared to mInt (53 %), which is relatively high compared to the negative control groups in the previous assays. The mutant D470A was included in the assay but could not be evaluated properly and is therefore not shown.

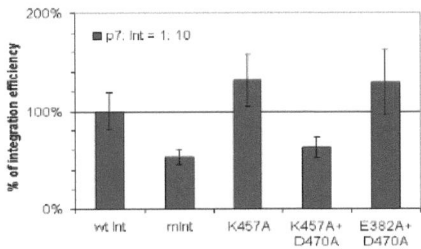

Figure 4.15. Integration efficiencies of selected integrase mutants in Hep1A cell line.
Two double mutants and one single mutant were compared to the control groups at a plasmid ratio p7:Int = 1:10 using a high integrase plasmid amount without additional stuffer plasmid. Experiments were carried out twice in triplicate.

Results

In summary, improved integration efficiencies were seen for several integrase point mutations in different mammalian cell lines. Integration efficiencies of the same mutant were not always consistent among different experiments. Increasing the plasmid dose showed increased colony formation in HeLa cells and 293 cells but not in HCT cells. Three different tendencies of recombination activity could be observed in HeLa cells (Figure 4.7). Synergistic effects with double mutants were not observed for every double mutant.

In the next step using a different setup addressing recombination efficiency, the integration efficiency of wt integrase should have been boosted by reducing its interaction with a recombination-decreasing protein.

4.5.5 Evaluation of integration specificity in DAXX siRNA transfected cells

PhiC31 integrase mediated integration efficiency has been demonstrated to be influenced by strong interaction with a cellular protein, a death domain associated protein, called DAXX. DAXX is present as an ubiquitary protein in human cells. Specific DAXX knock down with a DAXX-specific duplex RNA or small interfering RNA (siRNA) has been shown to increase recombination efficiency (Chen et al., 2006). During cellular processes with RNA interference (RNAi), specific double stranded RNA inhibits the expression of a particular gene in a sequence-specific manner (Fire et al., 1998).

Therefore, siRNA was used to reduce the production of DAXX protein. Cells transfected with DAXX specific siRNA were then used in a CFA to evaluate whether down regulation of DAXX has an effect on the integration efficiency of PhiC31 integrase. I expected an increase in integration efficiency due to a reduced interaction between inhibiting DAXX and the integrase.

HEK 293 cells were co-transfected with siRNA (DAXX specific and nonspecific) and plasmid DNA (wt Int or mInt encoding plasmids and substrate plasmid p7) (Section 3.1.5) and evaluated by RT PCR. The PCR efficiency and the relative down regulation of DAXX-specific siRNA compared to nonspecific siRNA were determined with the method "calculation by means of a standard curve" (Section 3.3.13) using obtained crossing points (cp). The cp value is frequently used as a measure to detect a certain PCR cycle, at which fluorescence significantly increases versus background fluorescence.

Differences in crossing points (cp) could be observed between the group treated with DAXX specific siRNA and the group treated with nonspecific siRNA (GAPDH specific) (Figure 4.16.A).The relative knock down of DAXX by DAXX-specific siRNA compared to nonspecific siRNA was determined to be 76 % (Figure 4.16.B). The down regulating effect of DAXX specific siRNA on DAXX expression was thereby confirmed.

Results

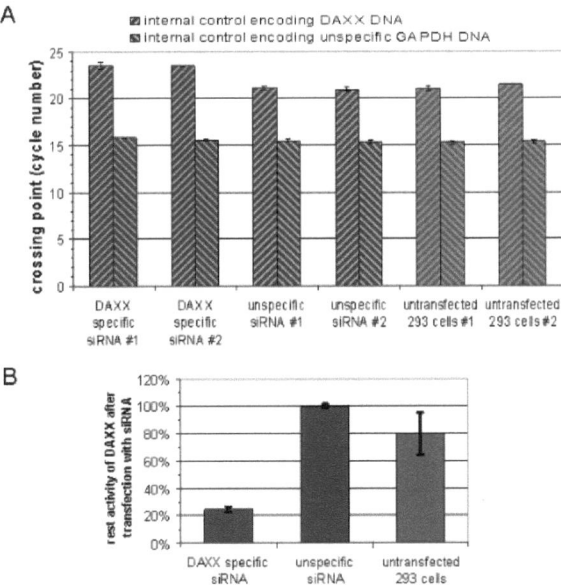

Figure 4.16. The relative knock down of DAXX by DAXX-specific siRNA in relation to nonspecific siRNA.
(A) Crossing point (cp) values after real-time PCR are shown for groups transfected with DAXX-specific siRNA, nonspecific siRNA and untransfected 293 cells. The left bar shows the cp values using the DAXX encoding plasmid as internal control, the right bar shows the cp values using the plasmid containing unspecific siRNA. (B) The rest activity of DAXX after specific and nonspecific siRNA-mediated knock down of DAXX is shown in siRNA transfected cells and untransfected control. Calculations are described in Section 3.3.13. DAXX siRNA samples are shown as left columns, reference values in the middle for nonspecific siRNA and right columns for untransfected cells.

Having tested and verified the down regulation of DAXX by DAXX-specific siRNA, integration efficiencies of active and inactive integrase versions were determined in context of siRNA transfected 293 cells. As a control for the siRNA transfected cells, nonspecific siRNA was used. The integration efficiencies of integrase in siRNA-transfected 293 cells are shown in Figure 4.17 below. No difference was observed between wt integrase mediated integration efficiency in cells without siRNA (black bar) and wt integrase mediated integration efficiency in cells transfected with DAXX specific siRNA. Integration efficiency of wt integrase transfected with nonspecific siRNA showed a 10 % decline. Likely, siRNA transfection 24 hours prior to integrase encoding plasmid transfection has no differential effect on integration efficiency between wt integrase and mInt. Considering the absolute number of colonies (between 80 and 100) in the recent assay (data not shown), the integration efficiency in this recombination assay is relatively

Results

low. Hence, the siRNA-specific down regulation of the DAXX protein has no increasing effect on integration efficiency in 293 cells using the present transfection conditions (plasmid ratios, choice of transfection reagent, FuGENE:DNA ratio, temporal distance between the transfections). Since the effect of siRNA on DAXX expression is only transient and lasts up to about four days, the suppressive effect of DAXX-specific siRNA onto the DAXX protein can only be sustained within the beginning of the selection process. DAXX is not suppressed for the entire duration of selection pressure on the cells. Hence, DAXX specific siRNA did not continuously suppress DAXX. The siRNA transfection could have been repeated two to three times during the selection, but were not due to these results. No further siRNA knock down experiments were carried out.

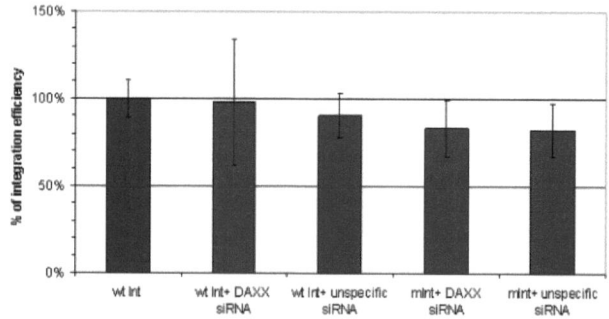

Figure 4.17. Integration efficiency of integrase in siRNA transfected 293 cells.
Integration efficiencies of wt Int and mInt are shown in 293 cells being siRNA transfected 24 hours prior to plasmid transfection. The DAXX specific siRNA was used as a reference, the GAPDH specific siRNA was used as control. HEK 293 cells were transfected with 100 pmoles siRNA by means of Lipofectamine 2000. One day later 2000 ng plasmid DNA with a molar plasmid ratio Int:p7 = 0.48:1 were transfected and another two days later, cells were splitted in 4×10^4 cells/ 10-cm dish. Transfections were carried out in triplicate.

The integration efficiency of PhiC31 integrase into mammalian cells has been evaluated in detail. As an additional approach, introduced in the next sections, the excision activity of PhiC31 integrase within intramolecular substrate DNA molecules was addressed using either chromosomal DNA or plasmid DNA. Both recombination events, integration and excision, have different applications in genome engineering but with discrepancies in efficiency.

Results

4.6 Integrase mediated excision in context of chromosomal DNA

An excision-based assay in chromosomal context of 293 cells was established using the reporter plasmid p-*attP*-polyA-*attB*-eGFP depicted in Figure 4.18. The reporter plasmid contains an expression cassette with a CMV promoter-driven eGFP gene. A transcriptional termination site (polyA signal), flanked on both sites by the native integrase recombination sites *attP* and *attB*, separates the transgene from its promoter. The removal of the polyA signal via integrase-mediated recombination leads to the fusion of the promoter to the GFP cDNA and consequently to expression. The expression of GFP as a measure for excision activity was assessed by flow cytometry using different transfection setups as described in Section 3.4.

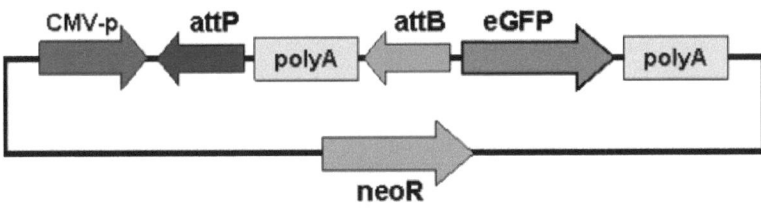

reporter plasmid p-attP-polyA-attB-eGFP (5574 bp)

Figure 4.18. The p-*attP*-polyA-*attB*-eGFP reporter plasmid.
The plasmid consists of a CMV promoter, followed by the integrase specific *attP* and *attB* sites, flanking a terminating polyA signal. The eGFP (enhanced green fluorescent protein) reporter gene downstream the *attB* site is turned on upon excision of the termination signal polyA. A neomycin resistance marker (bottom) allows for plasmid retention within the cell during G418 selection pressure. The plasmid was a great gift from Jia´s laboratory (Chen *et al.*, 2006).

First, the functionality of the GFP reporter plasmid was tested and confirmed upon transient transfection in a preliminary experiment. The GFP reporter plasmid was co-transfected with the wt Int encoding plasmid, with the mInt encoding plasmid or with the stuffer plasmid pBS in equal amounts (1000 ng per plasmid) in a 6-cm dish. Forty-eight hours post transfection the GFP expression from the transiently delivered plasmid was assessed by flow cytometry. Cells transfected with the wt Int encoding plasmid showed with a total fraction of 27 % fluorescent cells a 10 times higher GFP expression level than cells transfected with mInt encoding plasmid (2.7 %) or pBS (2.5 %).

Results

4.6.1 Establishment and evaluation of GFP reporter cell lines

HEK 293 cells were stably transfected with the GFP reporter plasmid (Figure 4.18) to establish reporter cell lines with conditional GFP gene activation in the presence of active integrase. These reporter cell lines conferring neomycin resistance cultured in D-MEM medium supplemented with G418 (500 µg/ml) were evaluated for GFP expression levels upon transfection of the Int plasmid.

The activation of the GFP gene within a stable chromosomal context depends not only on the recombination activity of the integrase to excise the polyA signal, but also on the targeted insertion site of the reporter plasmid within the chromosomal context, at which the reporter plasmid was inserted. Chromosomal features such as gene density and chromatin effects flanking the insertion site also influence the recombination and the subsequent GFP expression.

Therefore, each of the established reporter cell lines was individually transfected with the plasmid encoding the wt integrase to screen for a cell line which shows a rather high GFP expression potential in presence of active Int. Additionally, plasmid encoding mInt was solely transfected into each reporter cell line as a control to evaluate the function of the termination site blocking the reporter gene expression without active integrase. Cells were harvested 48 hours post integrase plasmid transfection and were analysed by flow cytometry to obtain the fraction of GFP expressing cells. This was done by gating the target cell population (10000 events), in which "living cells" were included, and among this cell fraction, the number of GFP expressing cells was determined.

Eleven reporter cell lines were evaluated for GFP expression (fraction of cells expressing GFP among all cells) in presence of active and inactive integrase. The ratio of GFP expression between cells transfected with wt Int and mInt encoding plasmids was measured in each cell line (Figure 4.19.A).

The GFP expression of wt Int transfected cells was about 2-4 times improved compared to cells transfected with mInt encoding plasmid. Active Int obtained between 6 % and 15 % GFP expression levels in 9 of the 11 analysed clones, except clone #20 and #45. The cell line based on clone #24 showed elevated GFP expression levels of 26.5 % and 13.6 % after transfection of plasmids encoding active or inactive versions of integrase. Taken the ratio between active and inactive integrase encoding plasmids into consideration, four reporter cell lines were selected for additional analysis (Figure 4.19.B). The selected reporter cell lines, termed according to the initially isolated numbered clone, were transfected with pBS as an additional negative control. GFP expression ratios between wt Int and mInt transfected cell lines resulted in ratios of 5.12:1 (for cell lines originated from clone #28), 3.95:1 (clone #31), 3:1 (clone #39) and 6:1 (clone #41), respectively (Figure 4.19.B). Based on these ratios, the cell line, originated from clone #28, was selected to test all constructed integrase mutants.

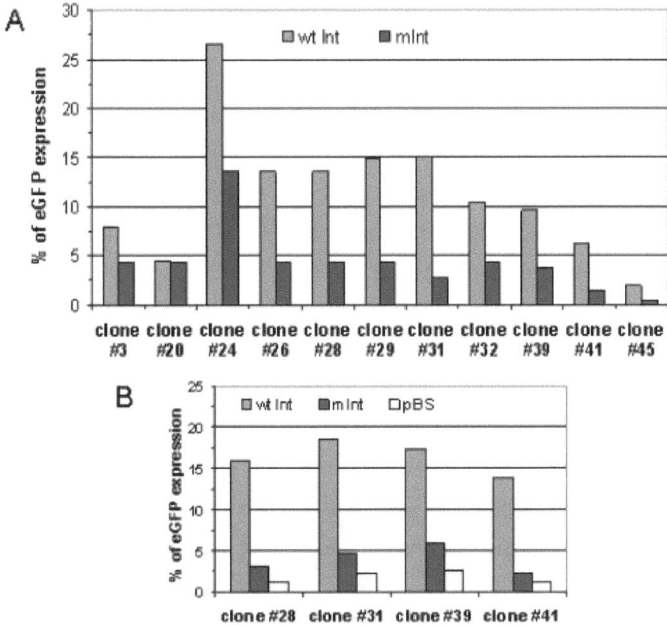

Figure 4.19. FACS analysis of various clones upon transfection of integrase plasmids.
HEK 293 cells were transfected with the plasmid shown in Figure 4.18 and selected in medium containing 500 µg/ml G418 containing medium. Generated cell lines were transiently transfected either with plasmids containing wt Int or mInt or with pBS plasmid. GFP expression was measured by flow cytometry via FACS as a measure of integrase mediated polyA excision. (A) Transfected cells were applied to FACS and eGFP expression was comparably detected between active (wt Int) and inactive (mInt) integrase. (B) Four promising clones (#28, #31, #39, and #41) show a high difference between wt Int and mInt mediated excision. Cell lines based on these four clones were transfected with wt Int or mInt encoding plasmid or with non-relevant stuffer DNA, respectively. Experiments were carried out in one replicate.

4.6.2 Evaluation of PhiC31 mutants in the eGFP reporter cell line #28

According to the GFP expression levels (Figure 4.19), the cell line based on clone #28 was chosen for further analysis of Int mutants. This selected cell line was co-transfected with an integrase encoding plasmid (wt Int, mInt or designed mutant) together with the plasmid pCR3mOrange. This plasmid encoding an "orange-fluorescent protein" can be used for compensation since eGFP and mOrange show different excitation maxima in their spectra. Co-transfections were carried out as described in Section 3.4.

Results

Plasmid-transfected cells were harvested and fluorescence-analysed by flow cytometry as described in Section 3.4. Cell properties were measured by FACS as described below. The first gating to target "living" cell populations reduced the size of the sample to at least 45 %, referred to 293 cells eGFP stable inactive ("eGFP inactive", left bars), which contains the "living" cells considered for further analysis, in which GFP is only active or expressed, when the integrase mediates recombination.

The gated population of "living" cells was analysed for orange-fluorescent cells (middle bars). The cell population transfected with both mOrange and eGFP fluorescence was considered to be the target population. Relative ratios of these GFP expression levels between wt Int transfected cells and cells transfected with Int mutants are shown in Figure 4.20. The FACS analysis displayed as dot plots and histograms is shown for the two control groups, wt Int and mInt, in Figure 4.20.A and 4.20.B. The gating of different cell populations according to their fluorescence properties was adjusted and the sizes of different cell fractions are graphically illustrated in Figure 4.20.C. The events (fluorescent properties) of individual cells were sorted by gating the individual cell fractions. The first gating (fraction of cells termed "gated 293 cells"; left bars) comprised 4833 and 5173 cell counts for the cells transfected with wt Int plasmid or mInt plasmids. These cell populations were considered as "healthy" or "living" and further analysed in terms of fluorescence properties. 2682 cells (or 55.5 %; middle bar) transfected with wt Int and mOrange plasmids showed mOrange expression. Among 5173 cells, 3019 cells (58.4 %; middle bar) being transfected with mInt and mOrange plasmids showed mOrange fluorescence. Within this cell fraction the positive control (wt Int) showed 1216 cells (right bar), which equals to 45.3 % of mOrange transfected cells, that express GFP. Hence, 1216 cells have taken up both plasmids. Cells transfected with mInt and mOrange obtained only 144 events (right bars), corresponding to 4.8 % of 3019 cells, which have taken up both plasmids.

Results

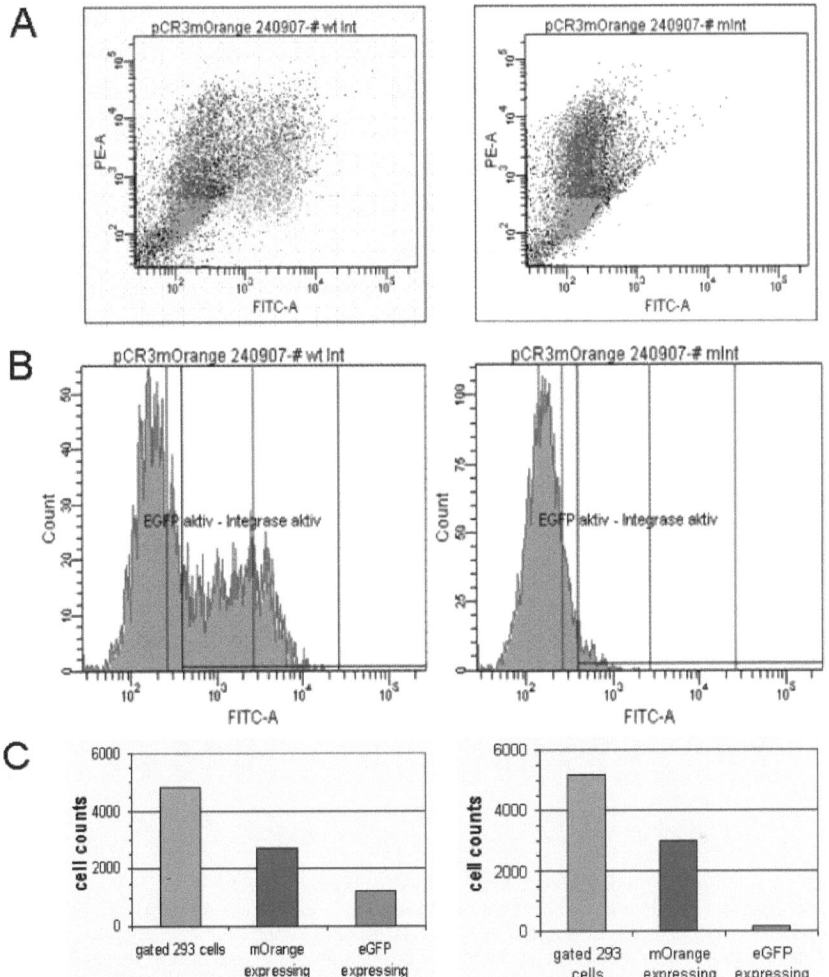

Figure 4.20. FACS analysis in reporter cell line #28.
The left panel shows raw data after analysis of the reporter cell line transfected with wt integrase. The right panel shows raw data after analysis of the reporter cell line transfected with mutated version (mInt). (A) Dot plots of the eGFP reporter cell line (#28) transfected show the distribution or sorting of single cells according to their fluorescence. Cells in the lower left corner represent the fraction being considered for fluorescence evaluation. Among this cell fraction, the cells above show orange fluorescent protein expression. Several dots among the population in the upper right area represent cells showing GFP expression among the cells fluorescing orange. This image is originally presented in colour for easier visualisation. (B) Histograms showing the frequency scale of eGFP fluorescing cells (counts) versus the eGFP intensity (x-axis). (C) The cell counts of each cell fraction or events per population are presented and sorted according to their fluorescence properties.

Results

The reporter cell line (#28) was transfected with plasmids containing Int mutants and pCR3mOrange. Forty-eight hours post transfection FACS analysis was performed and Int mutants were evaluated for their respective excision activity corresponding to corrected GFP expression (Figure 4.21). Data obtained by flow cytometry revealed eight mutants with similar excision activity compared to wt Int. The mutants E382A, D398A, E408A, R429A, E435A, R446A, K457A, and R461A resulted in excision activities comparable to wt Int. Two mutants R367A and D417A showed very low excision activity, even below that of mInt. Two other mutants, R380A and R393A, showed 40 % excision activity relative to wt Int. No integrase mutant with higher excision activity compared to wt Int was observed (Figure 4.21). Integrase-mediated excision within a chromosomal context could not be evaluated upon evaluation of the designed mutants using the present settings and conditions.

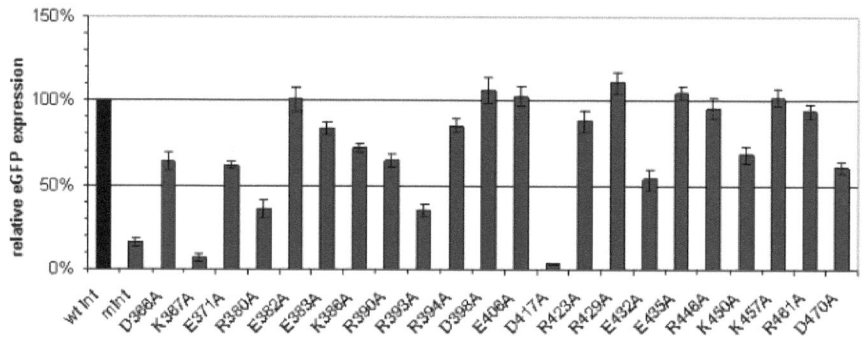

Figure 4.21. Integrase mediated excision activity of integrase mutants.
All 22 integrase mutants and two controls, wt Int and mInt, were evaluated for their excision activity in a genomic context. The reporter cell line (#28) was used. GFP expression values were normalised using the mOrange expressing cell population. GFP expression levels were detected by FACS analysis. wt Int was set to 100 %.

4.7 Integrase mediated excision within episomal plasmid DNA

4.7.1 An extrachromosomal assay to measure PhiC31 integrase mediated integration

In this experiment, the PhiC31 integrase-mediated recombination activity was analysed with a reporter on the plasmid pLucCR. pLucCR served as a substrate plasmid for excision and is depicted in Figure 4.22. The luciferase gene (yellow) represents the reporter, which is driven by a CMV promoter (red). An expression blocking site (polyA signal), which terminates promoter activity, is inserted between the promoter sequence and the luciferase gene. This termination sequence is flanked on both sites with the native PhiC31 integrase attachment sites. Plasmid construction was similar to the work introduced in Section 4.6 and Figure 4.18. Intramolecular recombination at the attachment sites is performed in the presence of PhiC31 integrase, which excises the terminating polyA signal. Consequently, the promoter and the coding sequence of the reporter gene are fused by hybrid site formation leading to luciferase expression. The principle of the excision-based recombination event is schematically illustrated in Figure 4.22. The integrase-mediated excision activity correlates with the luciferase expression levels, which are detected by a luminometer.

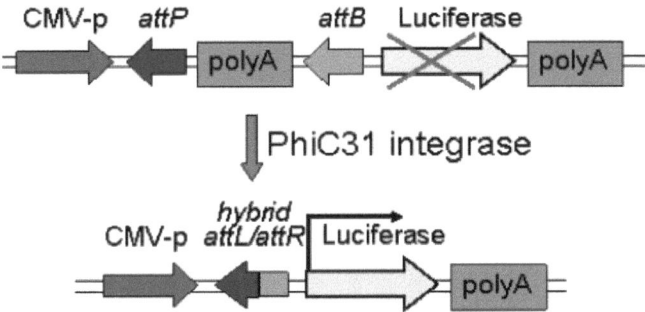

Figure 4.22. Schematic overview over luciferase expression after PhiC31 integrase-mediated polyA excision.
In presence of integrase the sequence between the *attB* and *attP*, sites are recombined out of the substrate plasmid pLucCR. Upon hybrid site formation (*attR* or *attL*), the CMV-promoter activates luciferase expression. CMV-p = CMV promoter, *attB* and *attP* = sequence specific attachment sites for PhiC31 integrase, polyA = polyadenylation signal to terminate translation and gene expression, respectively. Luciferase = reporter gene, hybrid *attR/L* = attachment sites after recombination of *attB* and *attP*.

The integrase-encoding plasmid, the substrate plasmid pLucCR, and an internal control plasmid encoding a *Renilla* luciferase, were transfected into 293 cells by transient triple transfections using different DNA ratios as described in Table 3.3. The two different types of luciferase enzymes, firefly luciferase and *Renilla* luciferase, present in different plasmids, could be activated under different conditions, and their activities, bioluminescent signals as relative light units (RLU), could be detected independently from each other. Bioluminescent signals by *Renilla* luciferase were used for normalisation of transfection. The corrected excision efficiency was calculated by forming the quotient between firefly luciferase light units and *Renilla* luciferase light units.

The data for the experiment shown in Figure 4.23 were based on a weight ratio of plasmids pLucCR:Int = 1:0.9 co-transfected with 60 ng (5 %) *Renilla* luciferase plasmid.

Results

The individual mutants were measured at least twice. When excision activities were highly different or values were not correctly detected, individual mutants were measured up to five times and final data were set in relation to the wt integrase excision activity and averaged. The overall luciferase assay showed diverse luciferase expression levels. Integrase mutants showed excision activities between 10 and 210 % compared to wt Int at 100 %. Five mutants, E371A, E383A, E406A, R423A, and K450A showed about twofold higher excision activity. Eleven mutants, D366A, E382A, K386A, R390A, R394A, D398A, R429A, E435A, R446A, K457A, and D470A obtained similar results as wt Int in range of 70-120 %. Six mutants, K367A, R380A, R393A, D417A, E432A, and R461A were inactive and showed between 5 % and 30 % excision activity (Figure 4.23).

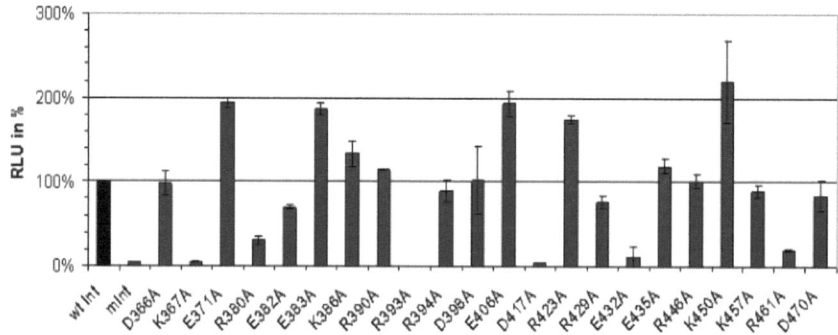

Figure 4.23. Excision activity of integrase mutants detected by a luciferase assay in 293 cells. Relative luciferase activity indicated as relative light units (RLU) corresponds to the excision activity of wt integrase (black bar) divided by the excision activity of integrase mutants. The mass ratio of plasmids was pLucCR:Int = 0.9:1. The luciferase experiments were performed in triplicate, and the results were averaged. Experiments were carried out between two and five times. The substrate plasmid is shown in Figure 4.22.

Moreover, to evaluate the influence of the integrase dose on the excision of the polyA signal, two different transfection ratios of pLucCR: Int = 1:1 and 1:5 were used. Forty-eight hours post transfection 293 cells were harvested and 10 µl lysed cells were measured for luciferase expression.

At a high amount of Int plasmids (pLucCR:Int = 1:5) three mutants, E371A, E383A, and K386A showed improved excision activities between about 130 % and 160 % compared to wt integrase. Ten mutants, E382A, R390A, E406A, D417A, R423A, R429A, E432A, E435A, R446A, and K457A, showed similar excision relative to luciferase expression levels of wt integrase (Figure 4.24). Five mutants, D366A, K367A, R380A, K450A, and D470A, showed excision activities below 60 %. The differences in excision activities of integrase mutants are due to different transfection setups (plasmid ratios) and conditions.

Results

The experiment shown in Figure 4.24 was performed in triplicate. Absolute RLU were higher using increased amounts of Int encoding plasmids (pLucCR:Int = 1:5), compared to the raw data from Figure 4.23.

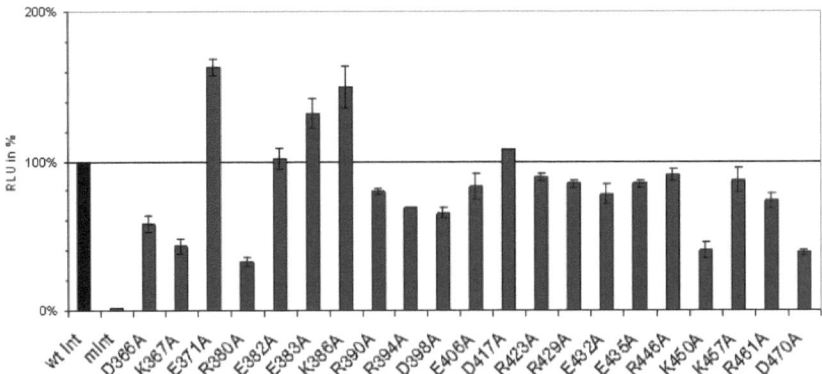

Figure 4.24. Excision activity of integrase mutants detected in a luciferase assay.
Relative luciferase activity indicated as relative light units (RLU) corresponds to the excision activity of wt integrase (left bar as cotrol) divided by the excision activity of integrase mutants. The plasmid ratio of pLucCR:Int = 1:5 was used in a total of 2000 ng plasmid transfected per well. Luciferase expression levels of wt Int were set to 100 %. Mean values are shown. Transfections were carried out in triplicate.

4.7.2 Construction of reporter plasmids containing favoured *pseudo attP* sites

By replacing the wt *attP* site with favoured *pseudo attP* sites the integrase-mediated specificity towards the *pseudo attP* sites could be addressed in a simple and fast excision based intramolecular assay. These preferentially targeted *pseudo attP* sites found at chromosomal positions 19q13.31, 12q22, and 2q11.2 also termed hot spots (hs), were identified among the most frequent sites of PhiC31 integrase mediated integration in human cell lines with an overall integration frequency between 2.9 % and 14.5 % (Ehrhardt et al., 2006). Insertion sites at chromosomal positions 2q11.2, 12q22, and 19q13.31 were found in 293 cells, HCT 116 cells and Huh7 cells, respectively. These three hot spots were targeted two times (2q11.2), ten times (12q22) and six times (19q13.31) in a total of 69 integration sites analysed (Ehrhardt et al., 2006). Two of these targeted sites, 2q11.2 in 293 cells and 19q13.31 in 293, D407 and HepG2 cells were also observed in an independent study (Chalberg et al., 2006). These three preferentially targeted insertion sites were selected to be used in intramolecular recombination assays in combination with the wt *attB* site since they were observed in two different studies, in different cell lines and on different chromosomes. These selected attachment sites should give a representative example as hot spots for PhiC31 integrase-mediated integration.

Results

The *pseudo attP* sites were obtained from the publications revealing PhiC31 integrase-mediated integration studies (Chalberg et al., 2006; Ehrhardt et al., 2006). Those query sequences were inserted into the sequence database using "nucleotide blast" as the BLAST program and the "human genomic plus transcript" as the selected database. The genomic DNA was isolated from 293 cells in order to be used as a PCR template to amplify these selected *pseudo attP* sites. The genomic sequences flanked on both sites of the particular hot spot comprising a length of about 350 bp were amplified by PCR using specific primers (Figure 4.25). For amplification of these *pseudo attP* sites, 400 ng eukaryotic DNA was used as template. PCR products were subcloned in pTOPO vector. After an additional PCR with oligonucleotides that included the restriction sites *AgeI* and *StuI*, the constructs were cloned into pLucCR replacing the native *attP* site. Modified substrate plasmids and the sequence of *pseudo attP* sites, which were inserted instead of the wt *attP* site, are schematically illustrated in Figure 4.26.A and 4.26.B, respectively.

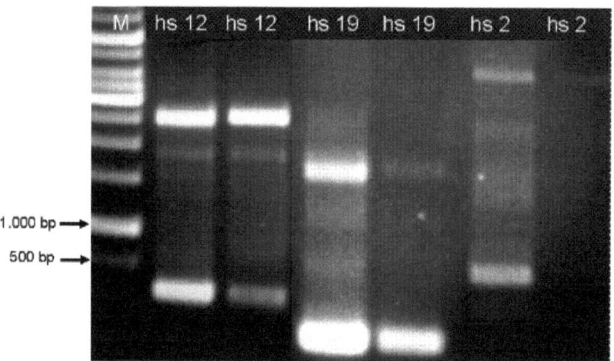

Figure 4.25. Analysis of amplified genomic hot spot DNA sequences with preferred PhiC31 integrase sites by agarose gel electrophoresis.
PCR was carried out using 400 ng genomic DNA isolated from 293 cells as template. Fragments with *pseudo attP* sites named according to their chromosomal position were amplified. The fragment hs12 (~330 bp) contains the *pseudo attP* site located at chromosomal position 12q22. hs19 (~290 bp) contains the *pseudo attP* site 19q13.31, and hs2 (~350 bp) contains the *pseudo attP* site 2q11.2. The particular *pseudo attP* sites were 39 bp long and surrounded by genomic DNA, obtained by nucleotide BLAST. PCR products of each hot spot were loaded onto two lanes and specific bands were gel-purified and subcloned. The remaining bands are referred to as nonspecific amplification products and were not considered. Two highlighted bands within the marker helped to identify the appropriate fragments. M= 1 kb DNA marker.

Results

A

Genomic location of integration site	Genomic sequence of integration site	Homology to wt *attP* site
native (wt) *attP* site	ACTGGGGTAACCT<u>TT</u>GAGTTCTCTCAGT	
pseudo attP site on chromosome 2q11.2	CCAGGGAAAAGCT<u>TC</u>AGTCTCTCCCTGG	15 / 28
pseudo attP site on chromosome 12q22	GTCCGGGGCGCCG<u>CT</u>CGGGTCTCCAGG	14 / 28
pseudo attP site on chromosome 19q13.31	CCACGGAATACCA<u>TA</u>GGGGTCACCAGGG	13 / 28

B

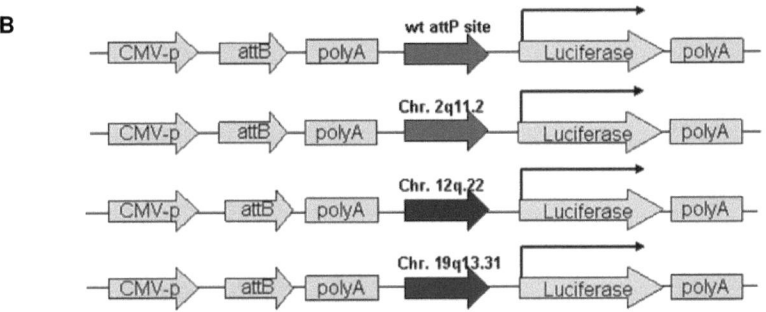

Figure 4.26. Plasmid constructs with wt *attP* site and three different *pseudo attP* sites.
The different *pseudo attP* sites were selected as preferred PhiC31 integrase attachment sites within different human cell lines (Chalberg et al., 2006; Ehrhardt et al., 2006). These particular *pseudo attP* sites were selected because they were found in two independent studies as favoured integration sites and were present in different cell lines. (A) Sequence alignments of three selected, preferentially targeted PhiC31 integrase specific *pseudo attP* sites. The minimal core sequence of wt *attP* site consists of a specific 5´-TT sequence. Sequence homology to wt *attP* is shown. (B) Excision constructs with different *pseudo attP* sites as shown in Figure 4.22. These constructs were used for luciferase assays. The wt *attP* site (on top) in the luciferase substrate plasmid was replaced either by the *pseudo attP* site from chromosome 2q11.2 from chromosome 12q22 or from chromosome 19q13.31. CMV promoter activates luciferase gene upon integrase-mediated excision of the polyA signal by binding to *attB* and wt or *pseudo attP* sites, respectively.

4.7.3 Evaluation of integrase mediated site-specific excision at *attB*/*pseudo attP*

In order to analyse *pseudo attP* site-specific excision by Int mutants, *pseudo attP* sites were derived from favoured genomic PhiC31 integrase target sites within the mammalian genome and inserted into the substrate plasmid. HEK 293 cells were co-transfected with integrase encoding plasmids and the substrate plasmid at a low Int dose pLuc:Int = 1:1 and at a high Int dose pLucCR:Int = 1:5. First, the *pseudo attP* site derived from chromosomal location 2q11.2 was analysed and luciferase results were shown in Figure 4.27.

Results

Comparing the results from the two experiments with different plasmid ratios showed that the relative fold change is higher in 9 of 21 mutants when a high amount of Int plasmid was transfected (right bars in Figure 4.27). Only mutants E382A, E383A, and K450A showed about 33 % higher excision activities at low dose of Int plasmid (pLucCR: Int = 1:1; left bars) compared to a plasmid ratio pLucCR:Int = 1:5 (right bars).

Twelve mutants, K367A, E371A, R380A, R386A, R390A, R394A, D398A, D417A, R432A, R429A, R448A and D470A, showed no significant difference in excision activity between a high and a low dose of Int plasmid. In both experiments, a few mutants showed improved excision compared to wt Int. At a plasmid ratio of 1:1 five Int mutants showed higher excision activity compared to wt Int (left bars > 100 %). At a plasmid ratio of pLucCR:Int = 1:5, nine mutants showed higher excision activity compared to wt Int (right >100 %). However, five of them showed only minor improvements up to 1.25-fold. The negative control mInt dropped only about 60 % compared to wt Int in both ratios.

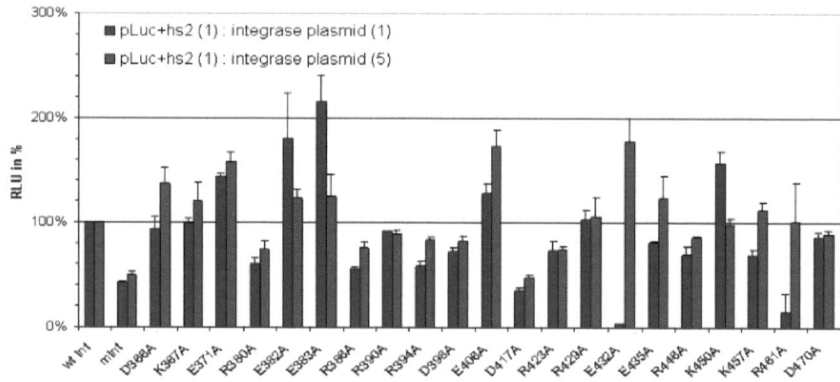

Figure 4.27. Excision activity of integrase mutants at *pseudo attP* site 2q.11.2 (abbreviated as hs2). Two assays were performed with a 1: 1 ratio (black bars) and a 1: 5 ratio (grey bars) of transfected plasmids pLucCR: Int plasmid. A total of 2000 ng plasmid DNA was transfected per well. Relative light units (RLU) of wt Int were set to 100 %. Experiments were carried out in triplicate.

Next, the *pseudo attP* site derived from 12q.22 was analysed for recombination (Figure 4.28). When Int plasmid was transfected fivefold in excess compared to substrate plasmid pLucCR, improved excision activity in 14 of 21 Int mutants was observed. Several Int mutants showed lower excision activity than wt Int in the context of *pseudo attP* site 12q.22. Only mutant E382A showed up to 1.5-fold higher excision activity than wt Int at low and high Int plasmid dose. The excision activity of mutant R429A increased 2.7-fold when increasing the integrase plasmid dose compared to substrate plasmid. Sixteen mutants showed lower excision activity between 20 % and 80 % compared to wt Int at both plasmid ratios. The excision activity of mInt dropped to 30 % and 45 %.

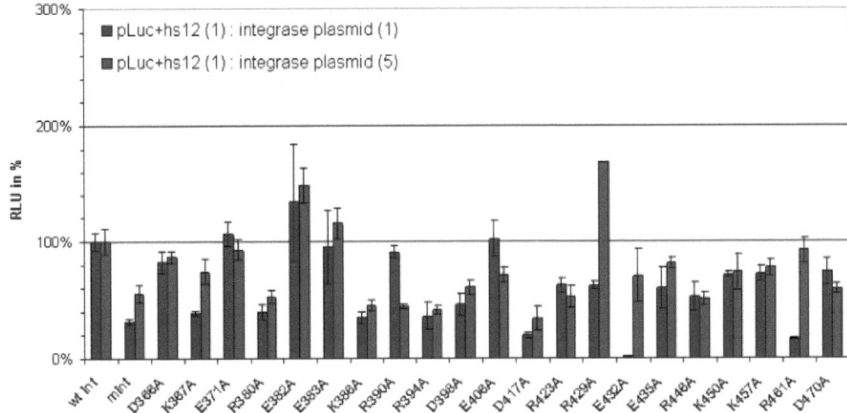

Figure 4.28. Excision activity of integrase mutants at *pseudo attP* site 12q.22 (hs12). Two assays were performed with a 1:1 ratio (black bars) and a 1:5 ratio (grey bars) of transfected plasmids pLucCR:Int plasmid. A total of 2000 ng plasmid DNA was transfected per well. Relative light units (RLU) of wt Int were set to 100 %. Experiments were carried out in triplicate.

Furthermore, the excision based assay was performed with substrate plasmid pLucCR containing *pseudo attP* site 19q.13.31 (Figure 4.29). The excision activity of integrase mutants at recombination target sites *attB* × *attP´* (19q13.31) confirmed that slightly higher excision activities were obtained when more Int plasmid was transfected. Only two mutants, K367A and K450A, showed a decrease of 30-40 % after enhancing Int plasmid dose. The mutant E371A obtained the highest excision efficiency with almost twofold increase at both plasmid ratios. Twelve mutants showed excision activities lower than wt Int-mediated excision. At a 1:1 ratio, the mutants E371A and K450A showed twofold and 1.4-fold higher excision activity than wt integrase. At high integrase plasmid dose, the mutants E371A, E382A, and E406A obtained increased excision activities between about 1.5- and twofold, compared to wt Int. The decrease of mInt compared to wt Int to 30 % and below at both low and high integrase plasmid dose proved the functionality of the assay.

Results

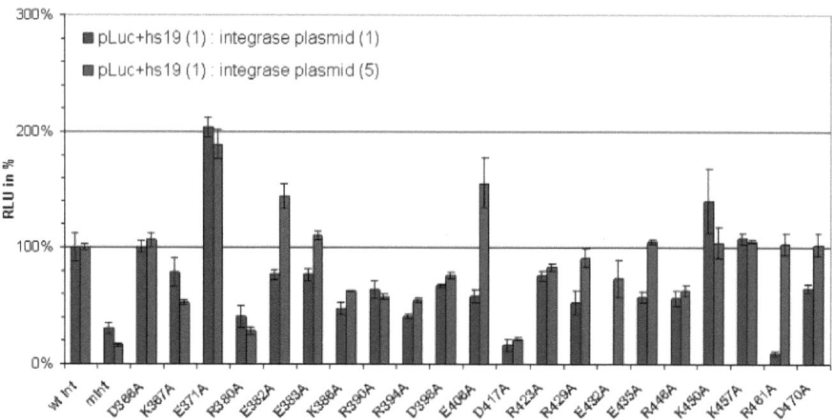

Figure 4.29. Excision activity of integrase mutants at *pseudo attP* site 19q13.31 (hs19). Two assays were performed with a 1:1 ratio (black bars) and a 1:5 ratio (grey bars) of transfected plasmids pLucCR: Int plasmid. A total of 2000 ng plasmid DNA was transfected per well. Relative light units (RLU) of wt Int were set to 100 %. Experiments were performed in triplicate.

Taken together, analysing the site-specific excision activity of integrase mutants in the context of different pairs of recombination recognition sites, wt *attB* × wt *attP*, wt *attB* × *attP´* (at chromosomal position 2q11.2), wt *attB* × *attP´* (12q.22), and wt *attB* × *attP´* (19q.13.31) in an intramolecular plasmid-based recombination assay showed high variations. Differences in recombination activity were not only observed among mutants recombining the same pair of recognition sites but also within single mutants at different *pseudo attP* sites. Individual mutants showed increased excision activity when the introduced *pseudo attP* sites were targeted. However, whether these mutants might integrate into the tested *pseudo attP* sites in context of genomic DNA as well, is highly influenced by variables like accessibility and potential preference for competing integration sites.

A summary showing relative recombination activities of integration and excision assays obtained from colony-forming assays and the GFP and luciferase reporter systems is presented in Table 4.2 below. Improved activities compared to wt integrase are presented in bold. The double mutants and D470A showed 1.5-fold to threefold improved integration efficiency in HeLa and HCT cells, and the mutants E371A, E382A, E383A, E406 and K450A showed between 1.3 and 2.2-fold excision activities in luciferase assays at different recognition sites.

Results

Table 4.2. Overview of PhiC31 integrase-mediated recombination activities in colony-forming assay, FACS and luciferase assay carried out *in vitro*. The recombination activity for each mutant is given in percentage compared to wt integrase (100 %). Plasmid ratios and plasmids are described.

Assay	Colony-forming assay								FACS	Luciferase assay							
cell line	HeLa				HCT	Huh7	293	Hep1A	#28	293							
figure	4.7	4.8	4.10	4.11	4.12	4.13	4.14	4.15	4.21	4.23	4.24	4.27		4.28		4.29	
plasmid names	substrate plasmid p7: Integrase plasmid									substrate plasmid pLucCR : integrase plasmid							
plasmid ratio	1:0.5	1:1	1:0.5	1:20	1:1	1:20	1:1	1:10		0.9:1	1:5	1:1	1:5	1:1	1:5	1:1	1:5
attB/ attP sites	all targeted attP sites within particular cell lines									attB × attP		attB × attP'		attB × attP'		attB × attP'	
position of attP										native		(2q11.2)		(12q.22)		(19q13.31)	
wt integrase	1.00	1.00	1.00	1.00	1.00	1.00	1.00	1.00	1.00	1.00	1.00	1.00	1.00	1.00	1.00	1.00	1.00
D366A	**1.22**								0.63	0.97	0.58	0.94	**1.37**	0.82	0.87	1.00	1.06
K367A	0.10								0.07	0.05	0.43	0.99	1.21	0.39	0.74	0.78	0.52
E371A	0.90								0.62	**1.94**	**1.63**	**1.44**	**1.58**	1.06	0.93	**2.04**	**1.89**
R380A	0.16								0.36	0.31	0.33	0.60	0.74	0.40	0.53	0.41	0.28
E382A	**1.36**		1.36						1.01	0.70	1.02	**1.80**	**1.23**	**1.34**	**1.49**	0.77	**1.44**
E383A	**1.28**								0.83	**1.87**	**1.32**	**2.16**	1.25	0.95	1.16	0.77	1.10
K386A	0.61								0.72	**1.34**	**1.50**	0.56	0.76	0.35	0.45	0.48	0.63
R390A	0.60								0.64	1.15	0.79	0.91	0.89	0.91	0.45	0.64	0.58
R393A	0.05								0.35								
R394A	0.13								0.85	0.89	0.69	0.59	0.83	0.36	0.42	0.41	0.55
D398A	0.76								1.07	1.02	0.65	0.73	0.83	0.46	0.61	0.67	0.76
E406A	0.86								1.03	**1.93**	0.83	**1.28**	**1.73**	1.02	0.71	0.58	**1.56**
D417A	0.03								0.04	0.04	1.08	0.35	0.47	0.20	0.34	0.16	0.22
R423A	0.81								0.88	**1.75**	0.90	0.74	0.74	0.63	0.53	0.76	0.83
R429A	0.60								1.11	0.77	0.85	1.03	1.06	0.62	**1.69**	0.53	0.91
E432A	0.09								0.53	0.12	0.78	0.03	**1.78**	0.01	0.71	0.00	0.73
E435A	0.77								1.05	**1.19**	0.85	0.81	1.23	0.60	0.82	0.57	1.05
R446A	1.04								0.95	1.01	0.91	0.69	0.86	0.53	0.51	0.56	0.63
K450A	0.04								0.68	**2.20**	0.40	**1.57**	0.98	0.72	0.74	**1.40**	1.04
K457A	**1.20**	1.66			1.95	0.7	1.17	1.3	1.02	0.89	0.87	0.69	1.12	0.72	0.78	1.08	1.05
R461A	**1.14**								0.94	0.20	0.73	0.15	1.01	0.16	0.92	0.09	1.02
D470A	**1.70**	1.47	1.45		0.42	1.05	1.09		0.60	0.84	0.38	0.86	0.89	0.74	0.60	0.65	1.02
E382A+D470A			1.70	3.3	1.27	2.33	1.44	1.3									
K457A+D470A			1.10	3.0	1.43	1.46	1.19	0.63									

Results

Based on these results, a few mutants were investigated for their detailed integration specificity in order to determine the specific location of integrase-mediated insertion within the genome of HCT cells.

4.8 Integration specificity of PhiC31 integrase in the genome of HCT cells

In order to characterise the integration profile of PhiC31 integrase within the human genome, the DNA sequences were analysed for integration sites, which were rescued from human cell lines. The human colon carcinoma-derived HCT cell line was used to determine PhiC31 integrase mediated integration specificity. HCT cells were co-transfected with the integrase expression plasmid pCS Int encoding either wt integrase or selected integrase mutants and with the substrate plasmid p7 carrying *attB* and a neomycin resistance gene. Single colonies, which were based on stable integrants due to G418 resistance, were selected for about two weeks and separately expanded to cell lines. From these cell lines, genomic DNA was isolated. Thereafter, integration sites were identified and analysed for chromosomal position by plasmid rescue (Section 3.2.4) (protocol adopted from Ehrhardt et al., 2006). The plasmid rescue, a convenient method to analyse genomic integration of extrachromosomal DNA, allows identification of both chromosomal sites flanking the hybrid *attL* and *attR* sites. Genomic integration is based on neomycin resistance over a period of more than two weeks. This technique however, might lead to artefacts resulting in only one rescued site.

The initial goal towards analysis of PhiC31 integrase specificity was to compare the integration profile of different PhiC31 integrase mutants within the genomic context of HCT derived cell lines. Approximately one hundred single clones were isolated and expanded to individual cell lines. Prior to isolation and cultivation of single clones, cells were transfected with the various versions of integrase encoding plasmid. Different numbers of clones (n) were expanded to cell lines, which were initially transfected with wt Int (n=21), mInt (n=8), mutant K457A (n=22), mutant D470A (n=22), double mutant K457A+ D470A (n=8), and double mutant E382A+ D470A (n=23). Rescued plasmids including p7 were first analysed by restriction digestion with *PstI* (Figure 3.4). The analytical digestion showed a characteristic fragment size of 1.6 kb and served as a preselection to identify rescued plasmids containing substrate plasmid sequence. Among the control-digested rescued plasmids, 50 samples were sequenced with specific sequencing primers and 44 rescued plasmids obtained quantitative sequencing results. These DNA sequences were then analysed for PhiC31 integrase-mediated integration events. DNA sequences, which were identified as derived from the human genome, were further analysed for the genetic content near the crossover site. The databases of bioinformatics websites were browsed for ORFs at or in the proximity of particular insertion sites. The human BLAST database or nucleotide BLAST at the NCBI website[9] and the UCSC Genome browser, with the BLAT Search Genome and database from UCSC Genome Bioinformatics website[10] were used (Cline and Ken, 2009; Kent, 2002; Kent et al,. 2002). Identified integration *pseudo attP* sites targeted by different versions of integrases, the respective sequencing primer, query versus genomic sequence size, the exact area and genomic location of each insertion event were presented. The genomic context near the integration site, which represented either a targeted gene or two flanking genes, if integration was intergenic, was also listed (Table 4.3). Each *pseudo attP* site was analysed for its chromosomal position and named according to the location, including the chromosome, the arm and band designation to differentiate between other *pseudo attP* sites. For example, the site 19q13.2 is located on chromosome 19 within the long arm "q", whereas the short arm is termed "p". Within the arm, the region "1", the particular band "3", and eventually a sub-band "2" describe the position in question, in this context an insertion site, in detail.

Eight different integration events could be indicated in the context of a human genome at locations within seven different chromosomes. The position 19q13.31 within the ORF of ZNF223 encoding a zinc finger was targeted twice in two independent events, mediated by wt integrase (clones #1 and #13.1). Another wt integrase performed recombination in an interagency region at position 20p13. Two insertion sites were determined at interagency positions, 13q21.2 and 4q28, by the integrase mutant D470A. The double mutant K457A+ D470A (clone #2) hit the largest among the human chromosomes at position 1q44 within ORF 71. The mutant D470A+ E382A (clone #23/1) hit two different chromosomal sites; more than two thirds of the queried sequences were found in chromosome X. Remaining sequences were identical to a small region (170-270 bp) in chromosome 6 within different arms. Three insertion sites at genomic location 19q13.2 and 20p13 and Xp11.22 could be identified by sequencing into both directions of the chromosome. Ideograms of all human chromosomes with the respective banding pattern and identified integration events (red arrows) are represented in Figure 4.30.

Remaining sequences did not reveal human genomic DNA. Most of the sequences obtained large parts of the transfected substrate plasmid p7 (data not shown). Therefore, these sequences could not be assigned to PhiC31 integrase mediated chromosomal integration.

---

[9] Nucleotide BLAST / BLAST human sequences at http://blast.ncbi.nlm.nih.gov/Blast.cgi
[10] Human BLAT search at http://genome.ucsc.edu/cgi-bin/hgBlat

Results

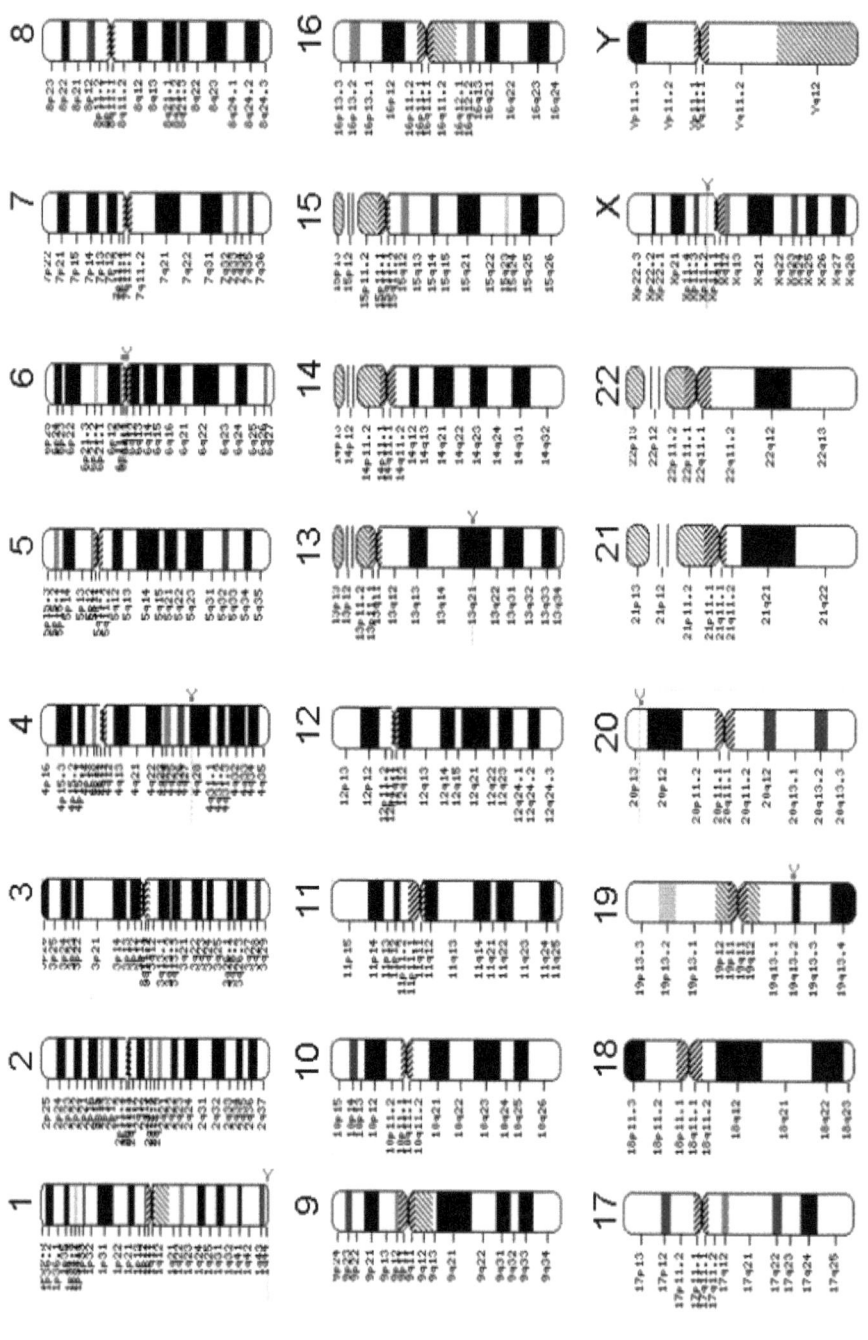

Figure 4.30. Positions of clonally rescued PhiC31 integration sites in the human genome. Chromosome ideograms are shown with integration sites marked as little arrows.

Table 4.3. Summary of rescued sites of PhiC31 integration in HCT cell line.

	Sequencing primer	Query size/ genomic sequence (bp)	Sequence start-sequence stop (bp)	Genomic location	If intronic, gene name	If intergenic, features flanking this part of subject sequence (intergenic sequence)
wt Int #1	attB-F	988/ 852	16832229-16833079	Chr. 19q13.31	ZNF223 mRNA	
	attB-R	991/ 851	16833079-16832229	Chr. 19q13.31	ZNF223 mRNA	
wt Int #3	attB-F	949/ 929	3719234-3720163	Chr. 20p13		CDC25B phosphatase (protooncogene) cell division cycle 25B isoform 1 and 3
	attB-R	908/ 457	3727808-3727352	Chr. 20p13		CDC25B phosphatase (protooncogene) cell division cycle 25B isoform 1 and 3
wt Int #13.1	attB-F	929/ 849	16832228-16833079	Chr. 19q13.31	ZNF223 mRNA	
D470A #2	attB-F	669/ 408 (266-665)	42769741-42769342	Chr. 13q21.2		647567 bp at 5' side: tudor domain containing 3 isoform 2; 195635 bp at 3' side: protocadherin 20
D470A #3/2	attB-R	589/589	50782528-50783119	Chr. 4q.28		602715 bp at 5' side: ankyrin repeat domain 50 2595 bp at 3' side: FAT tumor suppressor gene homolog 4
K457A+D470A #2	attB-F	943/ 893	246753004-246753906	Chr. 1q44,	C1 ORF 71	23574 bp at 5' side: mitochondrial TF B2 959 bp at 3' side: hypothetical protein LOC163882 isoform 2
D470A+E382A #23/1	attB-F	561/394 (1-397)	334782-334391	Chr. Xp11.22	shroom family #4	
		561/167 (394-561)		Chr. 6q11.1		323645 bp at 3' side: KH domain-containing, RNA-binding, signal transduction-associated protein 2
D470A+E382A #23/1	attB-R	948/ 601 (76-674)	334803-335403	Chr.Xp11.22	shroom family #4	
		948/277 (671-948)		Chr. 6p11.2		325094 bp at 3' side: KH domain-containing, RNA-binding, signal transduction-associated protein 2

Results

Among these few analysed insertion sites, no statistical predictions could be drawn. The chromosomal site 19q13.31 has also been found several times in previous studies analysing integrase specificity (Ehrhardt et al., 2006; Chalberg et al., 2006) and can rightly be assumed as a preferred integration site.

4.9 Human Factor IX (hFIX) expression upon plasmid delivery in murine liver

To analyse *in vivo* efficiencies of the different PhiC31 integrase mutants, plasmid DNA was transfected via hydrodynamic plasmid injection into the tail veins of mice. Two selected integrase mutants, K457A and D470A, which showed promising results *in vitro,* were compared to the control groups wt Int and mInt for integration into the genome of murine liver cells. Substrate plasmid pBS-*attB*-hFIX including the integrase specific wt *attB* site and the human blood coagulation Factor IX minigene (hFIXmg) was co-injected with either plasmid encoding wt Int, mInt, or with mutant K457A, D470A or K457A+ D470A, respectively (Table 4.4). Injected plasmids are shown in Figure 4.31.

At various time points, retro-orbital blood was collected. From the collected blood serum, liver toxicity was determined by detection of alanine aminotransferase (ALT) levels in both *in vivo* experiments (Figure 4.32). Additionally, hFIX levels in mouse serum were determined by enzyme-linked immunosorbent assay (ELISA).

In order to analyse the progression and the persistence of hFIX in murine liver tissue mice were treated at day 49 and day 71 post plasmid injection via intraperitoneal injection with carbon tetrachloride (CCl_4). CCl_4 is known to function as a liver-toxic substance and affected hepatocytes undergo necrosis. Application of 50 µl diluted CCl_4 leads to necrosis of at least 70 % of hepatic cells (Das et al., 2007). The original size of the CCl_4-damaged mouse liver was reconstituted by proliferating hepatocytes. Induction of the cell cycle leads to removal of episomal plasmids. Upon injection and DNA uptake in mouse liver hFIX plasmids persisted in the nucleus of hepatocytes either episomally or integrated into the genome. The hFIX concentration (in ng/ml) in individual mice of different groups (n = 3-5) were monitored at different time points in experiment #2 (Figure 4.33).

Results

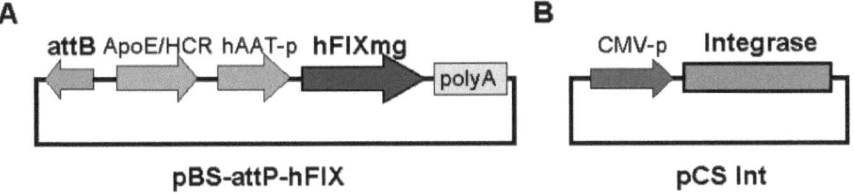

Figure 4.31. DNA sequences injected into female C57BL/6 mice.
The depicted plasmids were injected in different ratios into the tail vein of mice. (A) The pBS-*attB*-hFIX plasmid has already been used for hFIX delivery via injection (Ehrhardt et al., 2005). This plasmid contains the human blood coagulation Factor IX minigene (hFIXmg) as a reporter gene and the native *attB* site for PhiC31 integrase-mediated recombination. The expression cassette consists of different *cis*-acting regulatory sequences, the hFIXmg and the polyA sequence. The expression controlling sequences are the apolipoprotein E hepatic control region (ApoE/HCR), which further contains a matrix attachment region (MAR) and liver-specific enhancer elements, and the liver-specific human α1-antitrypsin promoter (hAAT-p). This promoter enhancer combination is introduced by Miao et al. (2000). (B) The plasmid pCS Int contains the CMV promoter (CMV-p) and the PhiC31 integrase gene, which is exemplary shown for all versions of integrase (wt Int, mInt and different integrase mutants) being used for tail vein injection.

Table 4.4. Experimental setup of two independent *in vivo* experiments.
Different characteristics for experiments #1 and #2 are shown. For experiment #1, 20 μg plasmid DNA were diluted in 2 ml 0.9 % NaCl solution and for experiment #2, 40 μg plasmid DNA were diluted in 1.8 ml 0.9 % NaCl solution. The appropriate amounts of diluted plasmids were injected via hydrodynamic tail vein injection in 6 seconds into the tail veins of mice. Plasmid ratios, individual plasmid amounts and numbers of mice per group are listed for both experiments.

In vivo experiment	Plasmid ratio pBS-*attB*-hFIX: Int	Amount of pBS-*attB*-hFIX plasmid	Amount of Int encoding plasmid and Int mutant	Mice evaluated per group
#1	1: 10	2 μg	18 μg wt Int	n= 3
#1	1: 10	2 μg	18 μg mInt	n= 3
#1	1: 10	2 μg	18 μg D470A	n= 5
#1	1: 10	2 μg	18 μg K457A+ D470A	n= 4
#1	Control mice uninjected			n= 2
#2	1: 1	20 μg	20 μg wt Int	n= 4
#2	1: 1	20 μg	20 μg mInt	n= 4
#2	1: 1	20 μg	20 μg K457A	n= 5
#2	1: 1	20 μg	20 μg D470A	n= 4
#2	Control mice uninjected			n.d.

The average ALT levels of the four treatment groups show expected progression over time. ALT levels were measured at day one post injection since ALT levels reach a peak shortly after plasmid injection. The toxicity profile relative to the ALT levels was monitored over time. ALT levels declined to normal range within eight days post injection and stayed normal (Figure 4.32A and 4.32.B). The normal ALT levels are in range between 10 and 40. However, they may vary greatly (Kaplan, 2002).

Results

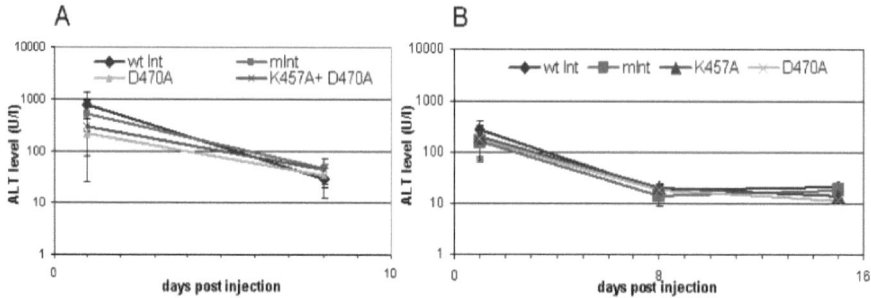

Figure 4.32. Surveillance of alanine transaminase (ALT) levels to evaluate liver damage.
ALT levels were determined in murine serum. ALT levels are represented in U/l (Units per litre). High-pressure tail vein injection was carried out at day 0. Two individual experiments, #1 and #2 as indicated in Table 4.4 were performed. (A) ALT expression levels of experiment #1 were detected at day 1 and 8 post injection. (B) ALT expression levels of experiment #2 were detected at day 1, day 8 and day 15 post injection.

The progression of hFIX concentration in different groups was determined by ELISA (Figure 4.33). hFIX levels increased about twofold between day 14 and day 28 post injection for all groups. Highest hFIX values could be obtained at day 42 post injection for wt Int (about 3500 ng/ml) and D470A (about 3000 ng/ml), mice co-injected with mutant K457A showed a decline to 1800 ng/ml and mInt to 900 ng/ml at day 42 post injection. The serum hFIX concentrations of wt Int (940 ng/ml) and integrase mutants K457A (957 ng/ml) and D470A (1061 ng/ml) were about 6 times higher than hFIX levels generated after injection with plasmids encoding mInt (140 ng/ml) at day 63 post injection.

After a second CCl_4 injection at day 71 post plasmid injection, hFIX serum levels even increased 1.33 times in wt Int transfected mice, 1.65 times in mice injected with mutant K457A, and 1.76 times in mice injected with mutant D470A.

At day 98 post injection, hFIX levels of the serum collected from wt Int or collected from the groups injected with integrase mutants dropped to a value of about 780 ng/ml, which is still about 25 times higher than the serum levels obtained from mice injected with mInt, approximately 30 ng/ml.

Results

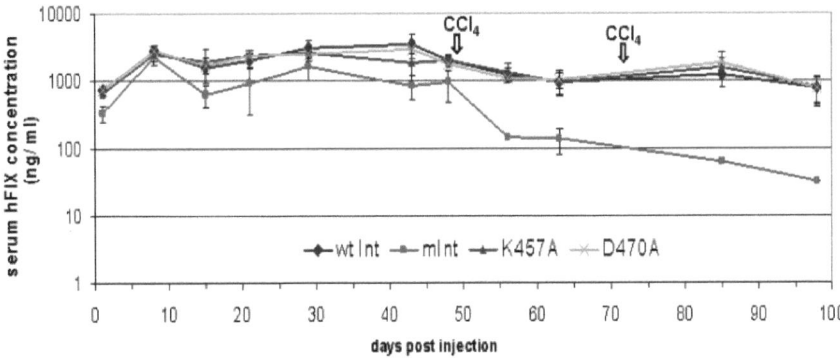

Figure 4.33. Human Factor IX (hFIX) expression levels *in vivo*.
Integrase encoding plasmids and hFIX encoding plasmids were co-injected via hydrodynamic delivery into the tail veins of female C57BL/6 mice according to the amounts in experiment #1 (Table 4.4). Blood serum samples from 4 different groups of mice, being injected with wt Int encoding plasmid, mInt, K457A mutant or D470A mutant, were analysed by ELISA for hFIX expression in murine liver at different time points. hFIX levels were averaged from all mice per group (n = 3-5). Fifty microlitres of liver toxic diluted CCl_4 (carbon tetrachlorid) were injected at day 49 and day 71 post injection to reduce the liver size and subsequently induce proliferation of hepatocytes and plasmid integration. hFIX progression is shown over a period of 14 weeks.

In order to address hFIX expression the progression of serum hFIX levels were detected starting at day 48 post injection. CCl_4 was injected at day 49 and day 71 post plasmid injection. The absolute (concentration in ng/ml) and the relative (in %) serum hFIX concentrations were presented for each mouse and group including percent values (Table 4.5). The serum hFIX concentrations from mice between the four groups but also in mice within the same group, at day 63 and 85 post injection had deviations higher than 100 %. Differences in serum concentrations of 100 % and higher were observed within all groups.

At day 85 post injection, the averaged hFIX serum level in mice injected with mutant K457A was 50 % higher than in mice injected with wt integrase. The averaged hFIX expression levels in mouse serum taken at the last time point of the experiment (day 98 post injection), however, were very similar to each other showing hFIX concentrations between 770 and 795 ng/ml in mouse serum collected from groups injected with wt Int and with the two Int mutants K457A and D470A.

Results

Table 4.5. Serum hFIX concentrations from different groups of mice injected with PhiC31 integrase plasmid and hFIXmg plasmid post initial CCl_4 injection are shown.
Supporting data to Figure 4.33 show the averaged hFIX concentration (in ng/ml) as absolute hFIX serum levels and the relative values in % from mice injected with wt Int, mInt, and mutants K457A and D470A in duplicates for each mouse per group. The values of hFIX concentrations are set to 100 % at day 48 post initial hydrodynamic plasmid delivery. From each mouse of the four different groups, blood samples were taken at day 48, 63, 85, and 98 post injections. The serum hFIX concentrations were determined by ELISA in ng/ml and the relative serum level was calculated for single mice within the same group and for the average in each group, based on the values at day 48 post injection. CCl_4 was injected at day 49 and 71 post initial injection to induce hepatic cell cycles. The number of mice in each group was n = 3-5. n.d. means not detected or below detection minimum for serum from mice injected with mInt.

days post injection	no. of mouse	Serum hFIX levels of individual mice per group and average in ng/ml and %							
		wt Int		mInt		K457A		D470A	
		ng/ml	%	ng/ml	%	ng/ml	%	ng/ml	%
day 48	#1	/		476	100 %	1918	100 %	1903	100 %
	#2	2029	100 %	1400	100 %	1983	100 %	1853	100 %
	#3	2230	100 %	963	100 %	2078	100 %	1482	100 %
	#4	2028	100 %	49	100 %	/		1910	100 %
	#5	1964	100 %	/		2299	100 %	/	
	#6	/		/		2016	100 %	/	
	average	2062,75	100 %	722	100 %	2058,8	100 %	1787	100 %
day 49		CCl_4 injection							
day 63	#1	/		105	22,1%	723	37,7%	1228	64,5%
	#2	620	30,6%	n.d.		543	27,4%	1075	58,0%
	#3	1213	54,4%	124	12,9%	971	46,7%	1265	85,4%
	#4	dead		n.d.		/		1085	56,8%
	#5	985	50,2%	/		1197	52,1%	/	
	#6	/		/		1349	66,9%	/	
	average	939,33	45,5%	114,5	15,9%	956,6	46,5%	1163,25	65,1%
day 71		CCl_4 injection							
day 85	#1	/		63	4,5%	1955	101,9%	1373	72,1%
	#2	716	35,3%	n.d.		n.d.	n.d.	2825	152,5%
	#3	1688	75,7%	n.d.		2004	96,4%	1430	96,5%
	#4	dead		n.d.		/		dead	
	#5	1344	68,4%			805	35,0%	/	
	#6	/				1551	76,9%	/	
	average	1249,33	60,6%	(63)	8,7%	1578,75	76,7%	1876	105,0%
day 98	#1	/		n.d.		1140	59,4%	n.d.	
	#2	422	20,8%	n.d.		n.d.		682	36,8%
	#3	1113	49,9%	32	3,3%	702	33,8%	859	58,0%
	#4	dead		n.d.		/		dead	
	#5	811	41,3%			543	23,6%	/	
	#6	/				n.d.			
	average	782	37,9%	(32)	4,4%	795	38,6%	770,5	43,1%

5. Summary

The PhiC31 integrase system represents an attractive tool for somatic integration of extrinsic genes or specific excision in applications such as therapeutic gene transfer or genetic engineering and transgenesis. In its most prominent role, the integrase mediates integration of plasmids carrying a therapeutic gene of interest and a specific *attB* site into *pseudo attP* sites. As a non-viral integrating vector, the PhiC31 integrase suffers from insufficient gene transfer and low integration efficiency. Additionally, aberrant events such as chromosomal rearrangements and deletions are known to occur at a given frequency of about 15 % within the host genome.

Therefore, the goal of this study was to first optimise the PhiC31 integrase-mediated integration efficiency and second to optimise the specificity by performing site-directed alanine scanning mutagenesis for the first time within the DNA binding domain.

In the course of this work, 22 single amino acid mutations were generated, which were then evaluated in antibiotic-selective integration assays in different human cell lines, based on co-transfection of integrase encoding plasmid and substrate plasmid with *attB* and neomycin resistance marker. Initial screening revealed one mutant with 1.7-fold higher integration efficiency. Upon optimisation of the plasmid ratio and with the creation of double mutants, the integration efficiency of two double mutants could be enhanced above threefold in HeLa cells. A nearly twofold improvement of one particular mutant could be achieved in Huh7 and in HCT cells, indicating that integration efficiency is cell line dependent.

Intramolecular plasmid-based excision assays were established leading to gene activation of the reporter. At native *attB* ×attP recombination sites, five mutants with about twofold enhanced integration efficiency compared to wt integrase were identified, whereas recombination activity in chromosomal context of a stably integrating GFP reporter cell line showed no improvement. Increasing integrase plasmid dose fivefold and replacing the native *attP* site within the substrate plasmid by three preferentially targeted *pseudo attP* sites found in mammalian genomes did not result in any improvement of the integrase-mediated excision activity of a polyA termination site beyond twofold.

To evaluate integration efficacy of PhiC31 integrase *in vivo* in murine liver a human coagulation Factor IX (hFIX) encoding substrate plasmid and the respective integrase encoding plasmid were co-injected via high-pressure tail vein injection into C57BL/6 mice. Two integrase mutants, K457A and D470A, showed a similar hFIX expression profile compared to wt integrase as determined by serum hFIX levels.

6. Discussion

6.1 Mutagenesis and other approaches to improve recombination of PhiC31 integrase

In order to improve non-viral gene transfer systems such as the PhiC31 integrase and the Sleeping Beauty (SB) transposase (Ivics et al., 1997), mutagenesis techniques to introduce mutations within their DNA sequences are widely used. Several studies aiming at identifying beneficial mutants by different screening assays have been carried out to improve integration efficiency and specificity on a protein level addressing the PhiC31 integrase (Sclimenti et al., 2001; Keravala et al., 2009) and the SB transposase (Geurts et al., 2003; Yant et al., 2004; Zayed et al., 2004; Mátés et al., 2009).

The mutagenesis screening approach used in this study was based on a widespread technique termed alanine scanning by replacing charged amino acids with alanine. For the first time, the DNA binding domain including amino acid positions 365-480 served as a template to generate 22 single mutants. Initial screening for integration efficiency of the integrase mutants resulted in an increase of 1.7-fold compared to the wild type integrase (Figure 4.7). A three- to fivefold enhancement in integration efficiency could be achieved in HeLa cells upon further improvements (Figure 4.11). The same method was used to change 95 single amino acids within the DNA binding domain of SB transposase resulting in 10 "hyperactive" mutants (10.5 %) with two- to fourfold enhanced transposition activity (Yant et al., 2004). Synergistic effects within the first generation transposases could boost the SB transposition up to a ninefold enhancement (Yant et al., 2004). Recent progress in improving stable gene transfer by a hyperactive version, created by a large-scale genetic screen of the SB showed 100-fold enhancement in efficiency (Mátés et al., 2009). Yet, such a breakthrough has not been achieved with the PhiC31 integrase system. When the PhiC31 integrase was mutated for the first time, the integration frequency and specificity could be improved two- to threefold at a favoured *pseudo attP* site on chromosome 8, detected by quantitative PCR (Sclimenti et al., 2001). In a study published recently, the N-terminal catalytic domain was screened for positive residues and among 43 mutants, only two (<5 %) showed up to 2.3-fold improvement in recombination efficiency in a "flipper assay". These mutants were created by error-prone PCR, included up to eight mutations and a 33 amino acid-long N-terminal fusion sequence, which was involved to improve recombination activity further (Keravala et al., 2009). The transfected substrate plasmid contained inverse oriented native *attB/attP* sites, flanking a CMV promoter,

Discussion

located upstream of an eGFP reporter gene. The recombination activity was determined by the inversion of the promoter, turning on eGFP, whose expression level was relative to the efficiency in the extrachromosomal "flipper assay" (Keravala et al., 2009). In concordance with the present study, their results demonstrate that double mutants do not consistently cause synergistic effects in respect to recombination efficiency. Increasing the number of mutations may result in a few beneficial mutants. Alternatively, specific amino acid residues can be replaced by others, which refers to a appropriate method, if more detailed structural would be available. However, limited data addressing structure, conformation, and enzymatic mechanism question this site-directed mutagenesis strategy.

Several other site-specific recombinases as lambda, *Cre*, and *FLP* and resolvases as gamma delta and Tn3 have been subjected to mutagenesis approaches addressing recombination and in particular DNA-binding, cleavage and ligation (Buchholz et al., 1998; Bankhead et al., 2003; Schwikardi and Dröge, 2000; Warren et al., 2005; Malanowska et al., 2009). Mutagenesis of lambda integrase and gamma delta resolvase can be designed more rationally and site-specific, since their putative protein structures are solved (Biswas et al., 2005; Li et al., 2005). The increasing number of studies addressing improvements of recombination and integration emphasise mutagenesis approaches still as a convenient method.

To increase integration efficiency further, the effect of enhancing integrase plasmid dose on colony formation was investigated. Dose dependent studies showed that the molar ratio of SB expressing plasmid to transposon containing plasmid affected SB-mediated integration. Overproduction inhibition consequently reduced transpositional activity at critical transposase concentrations (Geurts et al., 2003; Zayed et al., 2004), however N-terminal fusion of zinc fingers to a hyperactive SB transposase attenuated overproduction inhibition (Wilson et al., 2005). This adverse effect is not yet documented in PhiC31 integrase-mediated gene transfer. Thus, the effect of a twentyfold increase of integrase plasmid dose in respect to recombination efficiency was investigated. Although improvements could be achieved in integration efficiency of particular mutants caused by a higher plasmid dose in different cell lines (Figure 4.8 and 4.14), no general correlation between integrase plasmid dose and colony formation was detected. Addressing the influence of lowering the integrase plasmid dose showed a diminishing effect on recombination efficiency according to lower colony numbers (Figure 4.9). These findings are contradictory to recent data suggesting low dose of PhiC31 integrase to mediate considerable integration events in human cells (Wang et al., 2009). The interaction of extrinsic integrase with endogenous proteins as TTRAP (TRAF and TNF receptor

Discussion

associated protein) as presented by yeast two-hybrid and co-immunoprecipitation assays resulted in reduced effects on PhiC31 integrase recombination efficiency (Wang et al., 2009).

Another strategy was to combine improved single mutants to double mutants to achieve synergistic effects. Again, this method did not reveal a higher integration rate, however low synergistic effects were found in one double mutant in HeLa cells (Figure 4.10). The combination of both the double mutant and the plasmid dose increase shows a trend of higher activity slightly above threefold compared to wt integrase (Figure 4.11). Since further increase could not be achieved, the threshold of this experimental approach has potentially been reached. A triple mutant generated by three recombination-improved single mutants within the C-terminal domain of PhiC31 integrase also revealed a rather low synergistic effect in respect to recombination activity (Keravala et al., 2009). In the first mutagenesis study addressing SB, the combination of beneficial transposase mutants could be enhanced from initially fourfold by the factor of two (Yant et al., 2004). Large-scale mutagenesis by *in vitro* evolution using DNA shuffling performed on the genetic sequence of SB transposase even resulted in hyperactive transposases with up to hundredfold enhancement in efficiency (Mátés et al., 2009).

PhiC31 integrase fusion proteins with a nuclear localisation signal (NLS) (Kalderon et al., 1984) represent additional methods to enhance cell target specific delivery (Andreas et al., 2002). N-terminal tagged NLS to PhiC31 integrase reached almost the same recombination activity as NLS-*Cre* with an advancement of 1.74-fold compared to the native integrase (Andreas et al., 2002). Presumably, the addition of the NLS to our improved mutants might result in enhanced integration efficiency. The addition of NLS to the relatively large PhiC31 integrase is primarily associated with enhanced translocation through cellular membranes into the nucleus (Chen et al., 2006, Woodard et al., 2009). Surprisingly, the addition of the SV40 NLS had about fourfold decreasing effects towards integration efficiency *in vitro* in HeLa cells and *in vivo* upon hydrodynamic injection. Woodard et al. suggested that cell division was not required for integrase function in mouse liver and that nuclear localisation provided at most little benefits for PhiC31 integrase-mediated liver gene therapy (Woodard et al., 2009). Large fractions of enzymes in an unmodified state remain in the cytoplasm, since the nuclear membrane as pore complexes represent a limiting barrier for large proteins in size of 40 kDa and larger to enter the nucleus (Rolland, 2006; Bonner, 1978). A similar approach resulted in a more than threefold increase of targeted integration, when the combination mutants were fused with a 33 amino acid N-terminal sequence representing part of the bacterial/ mammalian promoter. This sequence is present in the native plasmid backbone and was additionally fused to integrase mutants (Keravala et al., 2009). A codon-optimisation towards a reduced number of CpG dinucleotides of the integrase in combination with the C-terminal addition of the simian virus 40 (SV40)-derived NLS showed twentyfold improvements in episomal recombination assays (Raymond and Soriano, 2007).

6.2 PhiC31 integration efficiency

In the selection-based integration assays, the efficiency of PhiC31 integrase to integrate plasmid DNA into the mammalian genome was determined relative to the quantity of G418-resistant colonies. The integration efficiency is predominantly influenced by the chromosomal context due to the intrinsic influence of the genomic DNA flanking the integration site, on the promoter and the transgene within the substrate plasmid. The choice and design of the substrate plasmid and in particular the promoter influences efficiency and transgene expression. The huge diversity of the chromosomal context within the genome derived from different cell lines and organs caused the inconsistent integration efficiency among different cell lines (Section 4.5). The finding, that chromatin conformation and accessibility were suggested to influence PhiC31 integrase pseudo site selection (Portlock and Calos, 2003) and transgene expression levels supports this observation. The frequency of integration by a related HIV integrase was shown to be influenced not only by the structure of the target site, but also by DNA curvature and flexibility (Pruss et al., 1994), thus proposing that features may also affect the PhiC31 integrase target site selection. Another parameter, which influences integration efficiency, refers to the transfection method (DNA delivery into the nucleus). Since equal transfection conditions via lipofection were assumed throughout the experiments, beneficial effects on plasmid delivery, e.g. addition of nuclear localisation signals (NLS), are not considered further within the context of this work. The overall integration efficiency upon lipofection-mediated transfection *in vitro* is generally low, but still higher than integration by homologous recombination (10^{-6}) for most mammalian cells (Vega, 1991). PhiC31 integrase-mediated integration naturally accomplished an approximate hundredfold improved recombination rate *in vitro* compared to random integration (Chalberg et al., 2006).

The constitution and the proficiency of the particular integrase likely play the most pivotal role in respect to integration efficiency. Since the crystal structure of the PhiC31 integrase has not been solved yet, homology modelling and secondary structure prediction (Yang

Discussion

and Steitz, 1995; Li et al., 2005) could support site-directed mutational analysis addressing recombination activity and in particular integration efficiency.

In the initial screen (Figure 4.7), five mutants showed a slight improvement in integration efficiency between 1.2- and 1.7-fold, assuming a mild functional preference for alanine instead of the large side chains of the charged amino residues (detailed structure in appendix). Single amino acid changes at these critical residues take certain effects on the overall enzymatic functionality e.g. by altering the charge and the conformation. However, replacing large side chains of amino acids, the methyl group of the small amino acid alanine does not necessarily lead to increased recombination efficiency.

Seven mutants showed a fivefold decline of integration efficiency suggesting structural or functional importance associated with the native charged amino acid positions (Figure 4.7). The small side group of alanine abolishes integration activity likely at these particular positions.

Ten mutants showed rather similar integration efficiency compared to wt integrase. Therefore, these amino acids were not critical to maintain the recombination activity at the level of wt integrase. Although the goal of this study targets to increase integration efficiency, it would be of great interest to determine, at which recombination step the mutant proteins are deficient, in order to gain more information and understanding about structure-function relationships. Initial experiments addressing synapsis, DNA cleavage and control of directionality during the PhiC31 integrase-mediated recombination reaction will lead to additional information about the enzymatic reaction (Thorpe et al., 2000; Smith et al., 2004).

According to the secondary structure prediction (Figure 1.5) the DNA binding domain comprises three repetitive β-sheets within position D366-R392 and a long α-helix comprising position between M413 and E480. A putative coiled-coil region, which is a common structural motif in proteins, in length of about 28 amino acids with a periodicity of four heptads forming several supercoiled α-helices has also been observed within the binding domain (Rowley et al., 2008; McEwan et al., 2009). In the first β-sheet comprising amino acid sequence 365-MDKLYC-370, two mutants (D366A and K367A) were created in this study, showing very diverse effects on integration efficiency. Removal of the acidic carboxy group (COOH) of aspartic acid (D) and replacing it with the methyl group gained slightly integration efficiency. The mutant K367A dropped integration activity about fivefold when the positive charge is removed. This might be due to the substitution of a large basic amino group by the small methyl group, which likely implies the importance of the polar side chain at that position to maintain hydrogen bonds between individual side chains of amino acids. The two mutants (E382A and E383A) showed both an approximate 1.3-fold increase. The residues are located between two connecting β-sheets. Mutations towards alanine at these positions might improve orientation and binding affinities between the second and the third β-sheet. The side-chain of the positively charged amino acid arginine (R) at position 380 might be involved to maintain the β-sheet structure of the second sheet, since alanine reduced activity dramatically. The third beta-sheet included two amino acids changed to alanine; K386A and R390A. The mutation effects resulted in a drop to 60 % integration efficiency compared to the wt integrase. Removal of positively charged side chains might consequently abolish the disruption of native interaction between the sheets, as indicated by the fivefold loss of recombination activity in the mutants R393A and R394A. These findings suggest that positively charged amino acids found in structures forming β-sheets support stronger binding to negatively charged DNA.

The most beneficial mutant D470A lies within the structural motif of a coiled-coil region between two alanine residues (Figure 4.5), in which four α-helices are coiled together in heptamers (Rowley et al., 2008; McEwan et al., 2009). By replacing the negatively charged side chain of aspartic acid (D) with alanine the protein interaction interface likely performs enhanced substrate recognition or improved α-helical propensity. The putative coiled-coil motif is strongly involved in substrate recognition and consequently in attBxattP synapsis (McEwan et al., 2009). The approximate tenfold decrease of mutants D417A and E432A suggests that the negatively charged sidechains of aspartic acid (D) and glutamic acid (E) are likely involved in protein active sites or cationic bonds rather than in DNA interaction. Whether or not the mutated residues in the double mutants are in close proximity in its active conformation could not be determined. Mutants showing improved efficiency were likely altered in specificity as well. However, it was not reviewed, if and when integration efficiency is associated with integration specificity, assuming that no obvious coherency exists.

6.3 Excision activity of PhiC31 integrase mutants

Besides PhiC31 integrase-based gene transfer associated with chromosomal integration, site-specific recombinases (SSR) are also capable of performing excision or inversion of DNA sequences when attachment sites are present within the same molecule, depending on their orientation to each other (Grindley et al., 2006; Sorrell and Kolb, 2005; Kolb, 2002).

Discussion

The plasmid recombination assay, used in this study was based on luciferase gene activation by the removal of the `expression blocking´ polyA sequence bringing luciferase cDNA in frame with the promoter. The intramolecular recombination assay carried out at native *attB/attP* sites revealed a few mutants with twofold improved recombination activity detected by luciferase expression (Figure 4.23 and 4.24). A time course of luciferase expression would nicely investigate and compare "long-term" expression potential of selected mutants using intramolecular recombination activity (Aneja et al., 2007). Evaluation of mutant derivatives for excision activity within extrachromosomal context excludes epigenetic effects as chromatin structure and modification, if integration would happen in regions with limited gene expression as heterochromatin.

In plasmids, the promoter activity is stable, assuming transient recombination assays relatively reliable. Excision activities of integrase mutants at preinserted plasmid constructs within chromosomal context of 293 cells however revealed no improvements in recombination activity, detected by GFP expression levels (Figure 4.21). This might be due to affecting chromatin position effects, which are capable to relieve transgene expression. This phenomenon could be avoided by insulating the integrated DNA with tandem copies of chicken β globin HS4 insulators, as used in transgenesis (Allen and Weeks, 2005).

Dimer interactions to form a synaptic complex with the substrate DNA (Rowley and Smith, 2008) might be preferentially achieved when target attachment sites are either present within one episomal molecule or present on two different molecules, as plasmid and genome. Although the excision principle is similar, results of both assays are rather ineligible to be compared, since the differences in terms of reporter gene detection (luciferase/ eGFP) and molecule (plasmid DNA/ genomic DNA) are crucial.

The excision-improved integrase mutants can be used in applications, where excision-based recombination is desired, e.g. removal of unwanted selectable markers. Excision-based recombination can be utilised for instance in transformed yeast cells, transgenic plants or embryonic stem cells, in animal transgenesis or conditional mutagenesis, to analyse specific gene functions. These applications have been dominated by related recombinases as *FLP* and *Cre* recombinase due to bidirectional reactions and their improved mechanistic and structural knowledge (Buchholz et al., 1996; Guo et al., 1997).

6.4 Specificity of PhiC31 integrase

Integration specificity gained great attention since aberrant events as chromosomal translocations and large deletions were associated with the PhiC31 integrase system *in*

Discussion

vitro at a frequency of about 15 % (Ehrhardt et al., 2006; Chalberg et al., 2006). Additional studies demonstrated that the PhiC31 integrase induces chromosomal aberrations and DNA damage response in primary human fibroblasts (Liu et al., 2006; Liu et al., 2009). Whether the cytotoxic effects were assigned to prolonged integrase expression or to high integrase doses, has not been investigated yet. Although the total number of potential integration sites was predicted to be in the range between 100 and 1000 due to a site-specific integration reaction (Chalberg et al., 2006), the risk of insertional mutagenesis still remains. In theory, the ultimate goal associated with PhiC31 integrase mediated gene transfer describes targeting the transgene expressing construct with high efficiency into a limited number of integration sites. Ideally, these sites are safe in terms of avoiding unwanted intrinsic side effects. Additionally, appropriate transgene expression levels should be achieved at these sites. Therefore, specificity analysis of integrase mutants using different methods were performed.

(1) Excision based assay evaluated specificity in terms of binding activity towards *attB* and *attP* sites within the same molecule. Excision frequencies were compared between native attachment sites *attB*×*attP* and 3 different combinations of *attB*×*attP'*. These *attP'* target sites were found to be preferentially targeted by the wild type integrase (Section 4.7.3, Figure 4.26).

(2) Integration assay based on plasmid rescue strategy aimed at detection of chromosomal sites of insertion (Figure 4.30, Table 4.3). Hence, this strategy produces more qualitative results. However, parameters such as design and size of substrate plasmid, selected restriction sites, and transfected cell lines bias the results (Ehrhardt et al., 2006; Chalberg et al., 2006). The transfer to *in vivo* analysis demands modifications in respect to choosing an appropriate marker gene.

Intramolecular excision studies based on recombination at *attB*×*attP* only compared the targeting frequency between *attP* and selected *attP'* sites. These sites might be preferentially targeted by the particular integrase mutants within genomic context as well. The evaluation of excision assays resulted in integrase mutants with about twofold improved specific recombination at selected *pseudo attP* sites. These slightly moderate improvements do not greatly support further applications of these mutants. Therefore, further trials to increase specificity need to be investigated, as shown in the next paragraph.

Results obtained from the plasmid rescue strategy yielded only very few integrase-mediated insertion events in HCT cells. Only the insertion site at position 19q13.31, which was found in two independent clones, has previously been identified as a preferred

Discussion

integration site in the two extensive PhiC31 integration studies to date (Ehrhardt et al., 2006; Chalberg et al., 2006). Most likely due to problems in plasmid rescue *per se* (frequent contamination with the substrate plasmid) and aberrant recombination events, insertion sites could not be detected properly. This system seems also to be error-prone, since artefacts of the rescue procedure might lead to the identification of only one single integration site (Chalberg et al., 2006). An alternative method to determine and analyse the integration events represents the PCR-based rescue by means of the "Genome walker" kit. This technique is based on linker ligation and nested PCR and determines only one site of the integration junction, being unable to detect rearrangements. Nested PCR could also introduce artefacts (Chalberg et al., 2006). The selection of restriction sites and the transfection method were influenced by the cell line to be analysed, due to the cell line's composition. The qualitative and specific analysis of integration sites helps to evaluate safety issues addressing the risk of insertional mutagenesis. Since the enzyme requires significant DNA sequence recognition (39 bp) for integration (Chalberg et al., 2006), much lower potential sites than retroviral or transposase-mediated integration systems are targeted (Mitchell et al., 2004; Schroeder et al., 2002; Wu et al., 2003; Yant et al., 2005). Additionally, severe adverse events in clinical trials were reported, in which retroviral vector-mediated insertion caused the proliferation of a proto-oncogene leading to leukaemia (Hacein Bey-Abina et al., 2003b). No similar adverse events in *in vivo* studies have been observed so far with non-viral integration systems. This makes PhiC31 integrase a potentially attractive tool for specific insertion into the genome, despite poor DNA delivery. Attempts to combine efficient delivery (high transduction efficiency) and site-specific PhiC31 integration were approached for the first time by the construction of a hybrid PhiC31 integrase adenovirus vector. The combination of both desirable properties leads to prolonged transgene expression *in vivo*, suggesting this system a new tool for gene therapy (Ehrhardt et al., 2007). Alternatives aiming at increasing specificity of SSRs represent the design of chimeric proteins as custom-made zinc finger nucleases (Alwin et al., 2005; Caroll, 2008) in combination with the PhiC31 integrase to target specifically DNA double strand breaks for gene targeting.

The precursor of the PhiC31 integrase referred to zinc finger fusion proteins has been the SB transposase, which was successfully linked to a zinc finger protein E2C, resulting in targeted integration into predefined sites in the human genome (Yant et al., 2007). These new generation transposases or chimeric proteins showed altered DNA-binding specificity and integration; however, controlled DNA transposition is still far from being realised perfectly.

6.5 Evaluation of integration efficiency in murine liver

To move further towards gene therapy trials, the PhiC31 integrase system was transferred to a mouse model. A simple, safe and effective method to transfer naked DNA molecules is the hydrodynamic tail vein injection of plasmids into the mouse (Zhang et al., 2004; Zhang et al., 1999; Liu et al., 1999). With a targeting frequency of at least 10 %, transgene expression via hydrodynamic delivery is preferentially found in hepatocytes (Herweijer and Wolff, 2003). One of the most appealing disease-related model systems describes gene transfer of human blood coagulation Factor IX (hFIX) into mouse liver to treat haemophilia. First, hydrodynamic injections with the respective integrase encoding plasmid and the donor plasmid pBS-*attB*-hFIX at a ratio of 10:1 were carried out (Table 4.4). This vector was designed and optimised for hepatic uptake by the inclusion of a liver-specific promoter and to treat haemophilia in gene therapy trials (Ehrhardt et al., 2003). According to improved integration efficiency in colony forming assays (CFAs), two mutants K457A+ D470A and D470A were transfected together with the substrate plasmid. In contrast to expectations, that surplus of integrase expressing plasmids compared to the donor plasmid might cause a positive impact on integration efficiency in murine liver. Almost all hFIX serum levels were below detection levels. Potential reasons for minimal hFIX expression levels might be insufficient amounts of transfected substrate plasmid (2 µg) in a total volume of 20 µg per mouse or a large plasmid dilution in 2 ml NaCl. During delivery through the blood stream into hepatocytes, injected plasmids are exposed to degrading enzymes as DNases. Only a fraction of proteins reached the target organ, which might have been to less for detection. Whether hFIX expression originated from extrachromosomally persisting plasmids or upon integration of the transgene was not determined, however appropriate expression at advanced time points are most likely associated with hFIX expression from integrated genes. The fate of the hFIX encoding plasmid (integration versus loss from the liver by out-dilution) is made within 48 hours post hydrodynamic injection (Woodard et al., 2009) and episomal plasmids are diluted out over time.

In the second *in vivo* experiment, 20 µg of both integrase plasmid and substrate plasmid were co-injected in a 1:1 ratio per mouse, adequate hFIX levels could be determined by ELISA over time (Figure 4.33, Table 4.5). Long-term and sustained transgene expression levels represent main prerequisites for successful gene therapy trials. Hepatocytes, as rather quiescent cells, simply cause a moderate cell turnover; however, the relatively harsh hydrodynamic delivery might initially increase proliferation of hepatocytes (Woodard et al.,

Discussion

2009). Transient expression and potential readministration to deliver therapeutic genes might be an alternative.

In murine liver, transient and acute toxicity was observed without any long lasting toxic effects (Miao et al., 2001). Whether the liver cells, which have taken up the hFIX expression plasmid, show different proliferation, must be speculated. The number of integration events per cell was discussed controversially within one group. One integration event per cell was proposed for wt PhiC31 integrase (Chalberg et al., 2006), whereas multiple integrations were also suggested for PhiC31 integrase mutants (Keravala et al., 2009), proposing elevated expression levels.

Comparing hFIX levels between my study and the recently published (Keravala et al., 2009) revealed slight differences in hFIX serum levels at day 85 post injection, despite liver toxic CCl_4 injection used in my study. Human factor IX levels revealed slightly higher concentrations of wt Int and mutants at day 85 post injection (Table 4.5) in the present experiment. Differences in hFIX long-term expression between both *in vivo* studies are mainly associated with differences in vector components of substrate and expression plasmids. Compared to the hFIX encoding plasmid in my study, the plasmid used in Keravala`s study is missing the second promoter ApoE including the hepatic locus control region (HCR). This vector additionally contains a Tn5 transposon in combination with a kanamycin resistance marker (Keravala et al., 2009). The differences in vector size and composition of features might lead to differences in expression levels. However, individual routes of delivery, localisation of the extrinsic protein, nuclear uptake, metabolism, and particular target sites have also particular influence on the expression levels. Repeated administration was shown to boost hFIX expression levels (Miao, 2005) and might even increase expression levels in the presented *in vivo* study. Immunofluorescence stainings on specific liver sections by using hFIX specific antibodies contributed to more detailed integration analysis and hFIX long-term expression profile (Keravala et al., 2009) and seemed appropriate to further evaluate and confirm long-term expression. The first study addressing therapeutic factor IX levels in mice could even detect prolonged survival of more than eight month of factor IX-expressing hepatocytes, suggesting no substantial immune response against integrase (Olivares et al., 2002).

Parameters to be investigated and optimised further in hepatic gene expression range from liver-specific α-1 antitrypsin (hAAT) promoters (Hafenrichter et al., 1994) over hepatic locus control regions (Ellis and Pannell, 2001) to enhancers and insulators, being involved in improvements of gene transfer into haemophilia mouse models (Miao et al., 2005).

6.6 Outlook and future perspectives

Aiming at increasing the recombination activity of the PhiC31 integrase further, several strategies in combination with the improved mutants might contribute to achieve this goal.

The construction and screening of additional mutants within the binding domain in combination with a detailed investigation of critical residues based on homology to related SSR, e.g. by applying a different mutational strategy than alanine scanning may gain additional mutants with improved recombination efficiency.

The determination of the crystal structure would definitely help to apply more rational and likely promising mutagenesis. A few common modifications on the protein level represent appropriate means to boost the PhiC31 integrase enzyme in particular. Combination of beneficial properties regarding delivery, nuclear uptake, and recombination will result in synergistic effects. A large-scale genetic screen involving gene shuffling, as successfully performed with the SB transposase (Mátés et al., 2009) will likely result in additional efficiency and specificity mutants. It would also be of interest to combine the construction of hybrid integrase adenovirus fusion proteins (Ehrhardt et al., 2007) with the new mutants found in the present study.

A next step might represent the generation of custom-built recombinases including a chimeric fusion of the integrase with a specific DNA binding domain e.g. a zinc finger nuclease that leads to specifically targeted double strand breaks near the site of the desired recombination (Caroll, 2008). Hence, the fusion construct can be used for gene targeting to any sequence in question, however this technique is still in an early state of research.

Improvements in vector development to prevent post integrative gene silencing and the steady decline in expression by exchanging promoters and polyA sites (Aneja et al., 2009) should also be introduced into the plasmids used in this study. The reduction or depletion of bacterial sequences or CpG islands (codon-optimisation) within the backbone vector of both the substrate and expression plasmids reduce silencing effects and can thereby prolong transgene expression. Approaches to guarantee transient presence of PhiC31 integrase or regulated integrase activity within a predetermined time window enhance the safety profile and are therefore worth performing (Sharma et al., 2008; Zhang et al., 2009). Binding assays as EMSA (electromobility shift assay) help to understand individual steps in the recombination reaction and evaluate further deficiency or proficiency of particular mutant proteins in combination with different attachment sites.

Discussion

Delivery and metabolism of foreign DNA in cells and animal models for the sake of researching and understanding human diseases better represents a high challenge for the future. Many efforts still need to be performed to improve persistence of transgenes and integration in a more controlled and safe manner in an existing complex biological network.

7. References

Abremski K. and Hoess R. (1984). Bacteriophage P1 site-specific recombination, purification and properties of the Cre recombinase protein. *J. Biol. Chem.* **259**, 1509-1514.

Aiuti A., Cattaneo F., Galimberti S., Benninghoff U., Cassani B., Callegaro L., Scaramuzza S., Andolfi G., Mirolo M., Brigida I., Tabucchi A., Carlucci F., Eibl M., Aker M., Slavin S., Al-Mousa H., Al Ghonaium A., Ferster A., Duppenthaler A., Notarangelo L., Wintergerst U., Buckley R.H., Bregni M., Marktel S., Valsecchi M.G., Rossi P., Ciceri F., Miniero R., Bordignon C. and Roncarolo M.G. (2009). Gene therapy for immunodeficiency due to adenosine deaminase deficiency. *N. Engl. J. Med.* **360**, 447-458.

Aiuti A., Slavin S., Aker M., Ficara F., Deola S., Mortellaro A., Morecki S., Andolfi G., Tabucchi A., Carlucci F., Marinello E., Cattaneo F., Vai S., Servida P., Miniero R., Roncarolo M.G. and Bordignon C. (2002). Correction of ADA-SCID by stem cell gene therapy combined with nonmyeloablative conditioning. *Science* **296**, 2410-2413.

Alba R., Bosch A. and Chillon M. (2005). Gutless adenovirus: last-generation adenovirus for gene therapy. *Gene Therapy* **12**, S18-S27.

Allen B.G. and Weeks D.L. (2009). Bacteriophage PhiC31 integrase mediated transgenesis in Xenopus laevis for protein expression at endogenous levels. Carroll D.J. (ed.) *Microinjection: Methods and Applications* **518**, 113-122.

Allen B.G. and Weeks D.L. (2005). Transgenic Xenopus laevis embryos can be generated using phiC31 integrase. *Nat. Methods.* **2**, 975-979.

Alwin S., Gere M.B., Guhl E., Effertz K., Barbas C.F.3rd, Segal D.J., Weitzman M.D. and Cathomen T. (2005). Custom zinc-finger nucleases for use in human cells. *Mol. Ther.* **12**, 610-617.

Andreas S., Schwenk F., Küter-Luks B., Faust N. and Kühn R. (2002). Enhanced efficiency through nuclear localization signal fusion on phage PhiC31-integrase: activity comparison with Cre and FLPe recombinase in mammalian cells. *Nucl. Acids Res.* **30**, 2299-2306.

Andrews B.J., Proteau G.A., Beatty L.G. and Sadowski, P.D. (1985). The FLP recombinase of the 2 micron circle DNA of yeast: interaction with its target sequences. *Cell* **40**, 795-803.

Aneja M.K., Geiger J., Imker R., Üzgün S., Kormann M., Hasenpusch G., Maucksch C. and Rudolph C. (2009). Optimization of Streptomyces bacteriophage φC31 integrase system to prevent post integrative gene silencing in pulmonary type II cells. *Exp. Mol. Med.* Sep 11. [Epub ahead of print]

Aneja M.K., Imker R. and Rudolph C. (2007). Phage phiC31 integrase-mediated genomic integration and long-term gene expression in the lung after nonviral gene delivery. *J. Gene Med.* **9**, 967-975.

References

Annenkov A.E., Daly G.M. and Chernajovsky Y. (2002). Highly efficient gene transfer into antigen-specific primary mouse lymphocytes with replication-deficient retrovirus expressing the 10A1 envelope protein. *J. Gene Med.* **4**, 133-140.

Arnold P.H., Blake D.G., Grindley N.D., Boocock M.R. and Stark W.M. (1999). Mutants of Tn3 resolvase which do not require accessory binding sites for recombination activity. *EMBO J.* **18**, 1407-1414.

Bainbridge J.W., Smith A.J., Barker S.S., Robbie S., Henderson R., Balaggan K., Viswanathan A., Holder G.E., Stockman A., Tyler N., Petersen-Jones S., Bhattacharya S.S., Thrasher A.J., Fitzke F.W., Carter B.J., Rubin G.S., Moore A.T. and Ali R.R. (2008). Effect of gene therapy on visual function in Leber`s congenital amaurosis. *N. Engl. J. Med.* **358**, 2231-2239.

Bankhead T.M., Etzel B.J., Wolven F., Bordenave S., Boldt J.L., Larsen T.A. and Segall A.M. (2003) Mutations at residues 282, 286, and 293 of phage lambda integrase excrt pathway-specific effects on synapsis and catalysis in recombination. *J. Bact.* **185**, 2653-2666.

Bateman J.R., Lee A.M. and Wu C.T. (2006). Site-specific transformation of Drosophila via PhiC31 integrase-mediated cassette exchange. *Genetics* **173**, 769-777.

Belteki G., Gertsenstein M., Ow D.W. and Nagy A. (2003). Site-specific cassette exchange and germline transmission with mouse ES cells expressing the PhiC31 integrase. *Nat. Biotechnol.* **21**, 321-324.

Belur L.R., Frandsen J.L., Dupuy A.J., Ingbar D.H., Largaespada D.A., Hackett P.B. and McIvor R.S. (2003). Gene insertion and long-term expression in lung mediated by the Sleeping Beauty transposon system. *Mol. Ther.* **3**, 501-507.

Bischof J. and Basler K. (2008). Recombinases and their use in gene activation, gene inactivation and transgenesis. *Methods Mol. Biol.* **420**, 175-195.

Bischof J., Maeda R.K., Hediger M., Karch F. and Basler K. (2007). An optimized transgenesis system for Drosophila using germ-line specific PhiC31 integrases. *Proc.Natl. Acad. Sci. USA.* **104**, 3312-3317.

Biswas T., Aihara H., Radman-Livaja M., Filman D. and Ellenberger T. (2005). A structural basis for allosteric control of DNA recombination by lambda integrase. *Nature* **435**, 1059-1066.

Blaas L., Musteanu M., Zenz R. Eferl R. and Casanova E. (2007). PhiC31-mediated cassette exchange into bacterial artificial chromosome. *Biotechniques* **43**, 659-660, 662, 664.

Bonner W.M. (1978). Proximity and accessibility studies of histones in nuclei and free nucleosomes . *Nucl Acids Res.* **5**, 71-85.

Bostanci A. (2002). Gene therapy. Blood test flags agent in death of Penn subject. *Science* **295**, 604-605.

References

Buchholz F., Angrand P.O. and Stewart A.F. (1998). Improved properties of FLP recombinase evolved by cycling mutagenesis. *Nat. Biotechnol.* **16**, 657-662.

Buchholz F., Ringrose L., Angrand P.O., Rossi F. and Stewart A.F. (1996). Different thermostabilities of FLP and Cre recombinases: implications for applied site-specific recombination. *Nucl. Acids Res.* **24**, 4256-4262.

Burke M.E., Arnold P.H., He J., Wenwieser S.V., Rowland S.J., Boocock M.R. and Stark W.M. (2004). Activating mutations of Tn3 resolvase marking interfaces important in recombination catalysis and its regulation. *Mol. Microbiol.* **51**, 937-948.

Bushman F.D. (2003). Targeting survival: integration site selection by retroviruses and LTR-retrotransposons. *Cell* **115**, 135-138.

Calos M.P. (2006). The PhiC31 integrase system for gene therapy. *Current Gene Ther.* **6**, 633-645.

Caroll D. (2008). Progress and prospects: Zinc-finger nucleases as gene therapy agents. *Gene Ther.* **15**, 1463-1468.

Cavazzana-Calvo M. and Fischer A. (2007). Gene therapy for severe combined immunodeficiency: are we there yet? *J. Clin. Invest.* **117**, 1456-1465.

Cavazzana-Calvo M., Hacein-Bey S., de Saint Basile G., Gross F., Yvon E., Nusbaum P., Selz F., Hue C., Certain S., Casanova J.L., Bousso P., Deist F.L. and Fischer A. (2000). Gene therapy of human severe combined immunodeficiency (SCID)-X1 disease. *Science* **288**, 669-672.

Chalberg T.W., Portlock J.L., Olivares E.C., Thyagarajan B. Kirby P.J., Hillman R.T., Hoelters J. and Calos M.P. (2006). Integration specificity of phage PhiC31 integrase in the human genome. *J. Mol.Biol.* **357**, 28-48.

Chater K.F. (1986). Streptomyces phages and their application to Streptomyces genetics. In the bacteria, Antibiotic-Producing Streptomyces, eds Queener, S.W. & Day, L.E., *Orlando, Florida, Acad. Press.* **9**, 119-158.

Check E. (2002). A tragic setback. *Nature* **420**, 116-118.

Chen L. and Woo S.L. (2008). Site-specific transgene integration in the human genome catalyzed by PhiBT1 phage integrase. *Human Gene Ther.* **19**, 143-151.

Chen L. and Woo S.L. (2005). Complete and persistent phenotypic correction of phenylketonuria in mice by site-specific genome integration of murine phenylalanine hydroxylase cDNA. *Proc. Natl. Acad. Sci. USA.* **102**, 15581-15586.

Chen J.Z., Ji C.N., Xu G.I., Pang R.Y., Yao J.H., Zhu H.Z., Xue J.I. and Jia W. (2006). DAXX interacts with phage PhiC31 integrase and inhibits recombination. *Nucl. Acids Res.* **34**, 6298-6304.

Chen Z.Y., He C.Y. and Kay M.A. (2005). Improved production and purification of minicircle DNA vector free of plasmid bacterial sequences and capable of persistent transgene expression *in vivo*. *Hum. Gene Ther.* **16**, 126-131.

References

Chou T.B. and Perrimon N. (1992). Use of a yeast site-specific recombinase to produce female germline chimeras in Drosophila. *Genetics* **131**, 643-653.

Chothia C. and Lesk A.M. (1986). The relation between the divergence of sequence and structure in proteins. *EMBO J.* **5**, 823-826.

Christiansen B., Brondsted L., Vogensen F.K. and Hammer K. (1996). A resolvase-like protein is required for the site-specific integration of the temperate lactococcal bacteriophage TP901-1. *J. Bacteriol.* **178**, 5164–5173.

Clamp M., Cuff J., Searle S.M. and Barton G.J. (2004). The Jalview Java Alignment Editor. *Bioinformatics* **20**, 426-427.

Cline M.S. and Kent W.J. (2009). Understanding genome browsing. *Nat. Biotechnol.* **27**, 153-155.

Cole A. (2008). Child in gene therapy programme develops leukaemia. *B.M.J.* **336**, 13.

Cole C., Barber J.D. and Barton G.J. (2008). The Jpred 3 secondary structure prediction server. *Nucl. Acid Res.* **36**, Web Server Issue W197-W201.

Collier L.S., Carlson C.M., Ravimohan S., Dupuy A.J and Largaespada D.A. (2005). Cancer gene discovery in solid tumors using transposon-based somatic mutagenesis in the mouse. *Nature* **436**, 272-276.

Conese M., Auriche C. and Ascenzioni F. (2004) Gene therapy progress and prospects: episomally maintained self-replicating systems. *Gene Ther.* **11**, 1735-1741.

Craig N., Craigie R., Gellert M. and Lambowitz A. (2002) *Eds., Mobile DNA II* (ASM), Washington DC, 272.

Cuff J.A. and Barton G.J. (2000). Application of multiple sequence alignment profiles to improve protein secondary structure prediction. *Proteins* **40**, 502-511.

Darquet A.M., Cameron B., Wils P., Scherman D. and Crouzet J. (1997). A new DNA vehicle for nonviral gene delivery: supercoiled minicircle. *Gene Ther.* **4**, 1341-1349.

Das R.K., Hossain S.U. and Bhattacharya S. (2007). Protective effect of diphenylmethyl selenocyanate against CCl_4-induced hepatic injury. *J. Appl. Toxicol.* **6**, 527-537.

Dong J.Y., Fan P.D. and Frizzell R.A. (1996). Quantitative analysis of the packaging capacity of recombinant adeno-associated virus. *Human Gene Ther.* **7**, 2101-2112.

Edelstein M.L., Abedi M.R. and Wixon J. (2007). Gene therapy clinical trials worldwide to 2007- an update. *J. Gene Med.* **9**, 833-842.

Ehrhardt A., Yant S.R. Ciering J.C., Xu H., Engler J.A. and Kay M.A. (2007). Somatic integration from an adenoviral hybrid vector into hot spot in mouse liver results in persistent transgene expression levels in vivo. *Gene Ther.* **15**, 146-156.

References

Ehrhardt A., Engler J.A., Xu H., Cherry A.M. and Kay M.A. (2006). Molecular Analysis of chromosomal rearrangements on mammalian cells after PhiC31-mediated integration. *Human Gene Ther.* **17**, 1077-1094.

Ehrhardt A., Xu H., Huang Z., Engler J.A. and Kay M.A. (2005). A direct comparison of two nonviral gene therapy vectors for somatic integration: *in vivo* evaluation of the bacteriophage integrase PhiC31 and the Sleeping Beauty transposase. *Mol. Ther.* **11**, 695-706.

Ehrhardt A., Peng P.D., Xu H., Meuse L. Kay M.A. (2003). Optimization of cis-acting elements for gene expression from nonviral vectors *in vivo*. *Human Gene Ther.* **14**, 215-225.

Ellis J. and Pannell D. (2001). The beta-globin locus control region versus gene therapy vectors: a struggle for expression. *Clin. Genet.* **59**, 17-24.

Enquist L.W., Kikuchi A. and Weisberg R.A. (1979). The role of λ integrase in integration and excision. Cold Spring Harbor Symp. *Quant. Biol.* **43**, 1115-1120.

Epstein A.L., Marconi P., Argnani R. and Manservigi R. (2005). HSV-1-derived recombinant and amplicon vectors for gene transfer and gene therapy. *Curr Gene Ther.* **5**, 445-458.

Fire A., Xu S., Montgomery M.K., Kostas S.A., Driver S.E. and Mello C.C. (1998). Potent and specific genetic interference by double-stranded RNA in Caenorhabditis elegans. *Nature* **391**, 806-811.

Friedmann T. and Roblin R. (1972). Gene therapy for human genetic disease? *Science* **175**, 949-955.

Gao X., Kim K.S. and Liu D. (2007). Nonviral gene delivery: What we know and what is next. *AAPS J.* **9**, E92-E104.

Garrison B.S., Yant S.R., Mikkelsen J.G. and Kay M.A. (2007). Postintegrative gene silencing within the Sleeping Beauty transposition system. *Mol. Cell. Biol.* **27**, 8824-8833.

Gaspar H.B., Parsley K.L., Howe S., King D., Gilmour K.C., Sinclair J., Brouns G., Schmidt M., von Kalle C., Barington T., Jakobsen M.A., Christensen H.O., Al Ghonaium A., White H.N., Smith J.L., Levinsky R.J., Ali R.R., Kinnon C. and Thrasher A.J. (2004). Gene therapy of X-linked severe combined immunodeficiency by use of a pseudotyped gammaretroviral vector. *Lancet* **364**, 2181-2187.

Geurts A.M., Yang Y., Clark K.J., Liu G., Cui Z., Dupuy A.J., Bell J.B., Largaespada D.A. and Hackett P.B. (2003). Gene transfer into genomes of human cells by the Sleeping Beauty transposon system. *Mol. Ther.* **8**, 108-117.

Glover D.J., Lipps H.J. and Jans D.A. (2005). Towards safe, non-viral therapeutic gene expression in humans. *Nat. Rev. Genet.* **6**, 299-310.

Golic K.G. and Lindquist S. (1989). The FLP recombinase of yeast catalyzes site-specific recombination in the Drosophila genome. *Cell* **59**, 499-509.

References

Grindley N.D., Whiteson K.L. and Rice P.A. (2006). Mechanisms of site-specific recombination. *Ann. Rev. Biochem.* **75**, 567–605.

Groth A.C., Fish M., Nusse R. and Calos M.P. (2004). Construction of transgenic Drosophila by using the site-specific integrase from phage phiC31. *Genetics* **166**, 1775-1782.

Groth A.C. and Calos M.P. (2004). Phage integrases: biology and applications. *J. Mol. Biol.* **335**, 667-678.

Groth A.C., Olivares E.C., Thyagarajan B. and Calos M.P. (2000). A phage intergase directs efficient site-specific integration in human cells. *Proc. Nat. Acad. Sci. USA* **97**, 5995-6000.

Guo F., Gopaul D.N., and Van Duyne G.D. (1997). Structure of Cre recombinase complexed with DNA in a site-specific recombination synapse. *Nature* **389**, 40-46.

Hacein-Bey-Abina S., Garrigue A., Wang G.P., Soulier J., Lim A., Morillon E., Clappier E., Caccavelli L., Delabesse E., Beldjord K., Asnafi V., MacIntyre E., Dal Cortivo L., Radford I., Brousse N., Sigaux F., Moshous D., Hauer J., Borkhardt A., Belohradsky B.H., Wintergerst U., Velez M.C., Leiva L., Sorensen R., Wulffraat N., Blanche S., Bushman F.D., Fischer A. and Cavazzana-Calvo M. (2008). Insertional oncogenesis in 4 patients after retrovirus-mediated gene therapy of SCID-X1. *J. Clin. Inv.* **118**, 3132-3143.

Hacein-Bey-Abina S., von Kalle C., Schmidt M., Le Deist F., Wulffraat N., McIntyre E., Radford I., Villeval J.L., Fraser C.C., Cavazzana-Calvo M. and Fischer A. (2003a). A serious adverse event after successful gene therapy for X-linked severe combined immunodeficiency. *N. Engl. J. Med.* **348**, 255-256.

Hacein-Bey-Abina S., von Kalle C., Schmidt M., McCormack M. P., Wulffraat N., Leboulch P., Lim A., Osborne C.S., Pawliuk R., Morillon E., Sorensen R., Forster A., Fraser P., Cohen J.I., ,de Saint Basile G., Alexander I., Wintergerst U., Frebourg T., Aurias A., Stoppa-Lyonnet D.,Romana S., Radford-Weiss I., Gross F., Valensi F., Delabesse E., Macintyre E., Sigaux F., Soulier J., Leiva L.E., Wissler M., Prinz C., Rabbitts T.H., Le Deist F., Fischer A. and Cavazzana-Calvo M. (2003b). LMO2-associated clonal T cell proliferation in two patients after gene therapy for SCID-X1. *Science* **302**, 415-419.

Hafenrichter D.G., Wu X., Rettinger S.D., Kennedy S.C., Flye M.W. and Ponder K.P. (1994). Quantitative evaluation of liver-specific promoters from retroviral vectors after *in vivo* transduction of hepatocytes. *Blood* **84**, 3394-3404.

Harris J.E., Chater K.F., Bruton C.J. and Piret J.M. (1983). The restriction mapping of c gene deletions in streptomyces bacteriophage PhiC31 and their use in cloning development. *Gene* **22**, 167-174.

Hatfull G.F. and Grindley N.D. (1988). Resolvases and invertases: a family of enzymes active in site-specific recombination. *In Genetic recombination*, American Society for Microbiology 357-396.

References

Heller L.C., Ugen K. and Heller R. (2005). Electroporation for targeted gene transfer. *Expert Opin Drug Deliv.* **2**, 255-268.

Herweijer H. and Wolff J.A. (2003). Progress and prospects: naked DNA gene transfer and therapy. *Gene Ther.* **10**, 453-458.

Hoess R., Abremski K., Irwin S., Kendall M. and Mack A. (1990). DNA specificity of the Cre recombinase resides in the 25 kDa carboxyl domain of the protein. *J. Mol. Biol.* **216**, 873-882.

Hollis R.P., Stoll S.M., Sclimenti C.R., Lin J., Chen-Tsai Y. and Calos, M.P. (2003). Phage integrases for the construction and manipulation of transgenic mammals. *Reprod. Biol. and Endocrinol.* **1**: 79.

Ivics Z., Hackett P.B., Plasterk R.H. and Izsvák Z. (1997). Molecular reconstruction of Sleeping Beauty, a Tc1-like transposon from fish, and its transposition in human cells. *Cell* **91**, 501-510.

Izsvák Z. and Ivics Z. (2004). Sleeping Beauty Transposition: Biology and Applications for Molecular Therapy. *Mol. Ther.* **9**, 147-156.

Izsvák Z., Khare D., Behlke J., Heinemann U., Plasterk R.H. and Ivics Z. (2002). Involvement of a bifunctional, paired-like DNA-binding domain and a transpositional enhancer in sleeping beauty transposition. *J. Biol. Chem.* **277**, 34581-34588.

Jaeger L. and Ehrhardt A. (2007) Emerging adenoviral vectors for stable correction of genetic disorders. *Curr. Gene Ther.* **7**, 272-283.

Kafri T., Morgan D., Krahl T., Sarvetnick N., Sherman L. and Verma I. (1998). Cellular immune response to adenoviral vector infected cells does not require *de novo* viral gene expression: implications for gene therapy. *Proc. Nat. Acad. Sci. USA* **95** 11377-11382.

Kalderon D., Roberts B.L., Richardson W.D. and Smith A.E. (1984). A short amino acid sequence able to specify nuclear location. *Cell* **39**, 499-509.

Kaplan M.M. (2002). Alanine aminotransferase levels: what`s normal? Editorial. *Annals of Internal Med.* **137**, 49-51.

Kelley L.A. and Sternberg M.J. (2009). Protein structure prediction on the Web: a case study using the Phyre server. *Nat. Protoc.* **4**, 363-371.

Kent W.J., Sugnet C.W., Furey T.S., Roskin K.M., Pringle T.H., Zahler A.M. and Haussler D. (2002). The human genome browser at UCSC. *Genome Res.* **12**, 996-1006.

Kent W.J. (2002). BLAT - the BLAST-like alignment tool. *Genome Res.* **12**, 656-64.

Keravala A., Lee S., Thyagarajan B., Olivares E.C., Grabosky V.E., Woodard L.E. and Calos M.P. (2009). Mutational derivatives of the PhiC31 integrase with increased efficiency and specificity. *Mol. Ther.* **17**, 112-120.

Kolb A.F. (2002). Genome engineering using site-specific recombinases. *Cloning Stem Cells* **4**, 65-80.

References

Kuhstoss S. and Rao R.N. (1991). Analysis of the integration function of the Streptomycete bacteriophage PhiC31. *J. Mol. Biol.* **222**, 897-908.

Levine A.J. (1987). Virus vector-mediated gene transfer *Microbiol. Sci.* **4**, 245-250.

Li W., Kamtekar S., Xiong Y., Sarkis G.J., Grindley N.D. and Steitz T.A. (2005). Structure of a synaptic γδ resolvase tetramer covalently linked to two cleaved DNAs. *Science* **309**, 1210-1215.

Liu J., Skjørringe T., Gjetting T. and Jensen T.G. (2009). PhiC31 integrase induces a DNA damage response and chromosomal rearrangements in human adult fibroblasts. *BCM Biotechnology* **9**: 31.

Liu J., Jeppesen I., Nielsen K. and Jensen T.G. (2006). PhiC31 integrase induces chromosomal aberrations in primary human fibroblasts. *Gene Ther.* **13**, 1188-1190.

Liu L., Mah C. and Fletcher B.S. (2006a). Sustained FVIII expression and phenotypic correction of hemophilia A in neonatal mice using an endothelial-targeted sleeping beauty transposon. *Mol. Ther.* **13**, 1006-1015.

Liu D., Ren T., Gao X. (2003). Cationic transfection lipids. *Curr Med Chem.* **10**, 1307-1315.

Liu F., Song Y.K. and Liu D. (1999). Hydrodynamics-based transfection in animals by systemic administration of plasmid DNA. Gene Ther 6, 1258–1266.

Lundstrom K. and Boulikas T. (2003). Viral and non-viral vectors in gene therapy: technology development and clinical trials. *Technol. Cancer Res. Treat.* **2**, 471-486.

Luo G., Ivics Z., Izsvák S. and Bradley A. (1998). Chromosomal transposition of a Tc1/mariner-like element in mouse embryonic stem cells. *Proc. Nat. Acad. Sci. USA* **95**, 10769-10773.

Lupas A. (1997). Predicting coiled-coil regions in proteins. *Curr. Opin. Struct. Biol.* **7**, 388-393.

Maguire A.M., Simonelli F., Pierce E.A., Pugh E.N. Jr., Mingozzi F., Bennicelli J., Banfi S., Marshall K.A., Testa F., Surace E.M., Rossi S., Lyubarsky A., Arruda V.R., Konkle B., Stone E., Sun J., Jacobs J., Dell'Osso L., Hertle R., Ma J.X., Redmond T.M., Zhu X., Hauck B., Zelenaia O., Shindler K.S., Maguire M.G., Wright J.F., Volpe N.J., McDonnell J.W., Auricchio A., High K.A. and Bennett J. (2008). Safety and efficacy of gene transfer for Leber's congenital amaurosis. *N. Engl. J. Med.* **358**, 2240-2248.

Mairhofer J. and Grabherr R. (2008). Rational vector design for efficient non-viral gene delivery: challenges facing the use of plasmid DNA. *Mol. Biotechnol.* **39**, 97-104.

Malanowska K., Cioni J., Swalla B.M., Salyers A. and Gardner J.F. (2009). Mutational analysis and homology-based modelling of the IntDot core-binding domain. *J. Bact.* **191**, 2330-2339.

References

Marshall E. (1999). Gene therapy death prompts review of adenovirus vector. *Science* **286**, 2244-2245.

Mátés L., Chuah M.K., Belay E., Jerchow B., Manoj N., Acosta-Sanchez A., Grzela D.P., Schmitt A., Becker K., Matrai J., Ma L., Samara-Kuko E., Gysemans C., Pryputniewicz D., Miskey C., Fletcher B., VandenDriessche T., Ivics Z. and Izsvák Z. (2009). Molecular evolution of a novel hyperactive sleeping beauty transposase enables robust stable gene transfer in vertebrates. *Nat. Genet.* **41**, 753-761.

Mazda O., Satoh E., Yasutomi K. and Imanishi J. (1997). Extremely efficient gene transfection into lympho-hematopoietic cell lines by Epstein-Barr virus-based vectors. *J. Immunol. Methods* **204**, 143-151.

McEwan A.R., Rowley P.A. and Smith M.C. (2009). DNA binding and synapsis by the large C-terminal domain of PhiC31 integrase. *Nucl. Acid Res.* **37**, 4764-4773.

Miao C.H. (2005). A novel gene expression system: non-viral gene transfer for hemophilia as model systems. *Adv. in Genetics* **54**, 143-177.

Miao C.H., Thompson A.R., Loeb K. and Ye X. (2001). Long-term and therapeutic-level hepatic gene expression of human factor IX after naked plasmid transfer *in vivo*. *Mol. Ther.* **3**, 947-957.

Miao C.H., Ohashi K., Patijn G.A., Meuse L., Ye X., Thompson A.R. and Kay M.A. (2000). Inclusion of the hepatic locus control region, an intron, and untranslated region increases and stabilizes hepatic factor IX gene expression in vivo but not in vitro. *Mol. Ther.* **1**, 522-532.

Miao C.H., Snyder R.O., Schowalter D.B., Patijn G.A., Donahue B., Winther B. and Kay M.A. (1998). The kinetics of rAAV integration in the liver. *Nat Genet.* **19**, 13-15.

Mikkelsen J.G., Yant S.R., Meuse L., Huang Z., Xu H. and Kay M.A. (2003). Helper-independent Sleeping Beauty transposon-transposase vectors for efficient nonviral gene delivery and persistent gene expression *in vivo*. *Mol. Ther.* **8**, 654-665.

Mitchell R.S., Beitzel B.F., Schroeder A.R., Shinn P., Chen H., Berry C., Ecker J.R. and Bushman F.D. (2004). Retroviral DNA integration: ASLV, HIV, and MLV show distinct target site preferences. *PLOS Biol.* **2**, e234 (1127-1137).

Murley L.L. and Grindley N.D. (1998). Architecture of the gamma delta resolvase synaptosome: oriented heterodimers identity interactions essential for synapsis and recombination. *Cell* **95**, 553-562.

Nakai H., Montini E., Fuess S., Storm T.A., Meuse L., Finegold M., Grompe M. and Kay M.A. (2003a). Helper-independent and AAV-ITR-independent chromosomal integration of double-stranded linear DNA vectors in mice. *Mol Ther.* **1**, 101-111.

Nakai H., Montini E., Fuess S., Storm T.A., Grompe M. and Kay M.A. (2003b). AAV serotype 2 vectors preferentially integrate into active genes in mice. *Nat. Genetics* **34**, 279-302.

References

Nakai H., Yant S.R., Storm T.A., Fuess S., Meuse L. and Kay M.A. (2001). Extrachromosomal recombinant adeno-associated virus vector genomes are primarily responsible for stable liver transduction *in vivo*. *J. Virol.* **75**, 6969-6976.

Nakai H., Iwaki Y., Kay M.A. and Couto L.B. (1999). Isolation of recombinant adeno-associated virus vector-cellular DNA junctions from mouse liver. *J Virol.* **73**, 5438-5447.

Nehlsen K., Broll S. and Bode J. (2006). Replicating minicircles: Generation of nonviral episomes for the efficient modification of dividing cells. *Gene Ther. Mol. Biol.* **10**, 233-244.

Nielsen T.O. (1997). Human germline gene therapy. (MJM) *McGill Journal of Medicine* **3**, 126-132.

Niidome T. and Huang L. (2002). Gene therapy progress and prospects: nonviral vectors. *Gene Ther.* **9**, 1647-1652.

Olivares E.C., Hollis R.P., Chalberg T.W., Meuse, L., Kay M.A. and Calos, M.P. (2002). Site-specific genomic integration produces therapeutic factor IX levels in mice. *Nat. Biotechnol.* **20**, 1124-1128.

Olivares E.C., Hollis R.P. and Calos M.P. (2001). Phage R4 integrase mediates site-specific integration in human cells. *Gene* **278**, 167-176.

Ortiz-Urda S., Thyagarajan B., Keene D., Lin Q., Fang M., Calos M.P. and Khavari P.A. (2002). Stable nonviral genetic correction of inherited human skin disease. *Nat. Med.* **8**, 1166-1170.

Palmer D. and Ng P. (2003). Improved system for helper-dependent adenoviral vector production. *Mol. Ther.* **8**, 846-852.

Palmer J.A., Branston R.H., Lilley C.E., Robinson M.J., Groutsi F., Smith J., Latchman D.S. and Coffin R.S. (2000). Development and optimization of herpes simplex virus vectors for multiple long-term gene delivery to the peripheral nervous system. *J. Virol.* **74**, 5604-5618.

Piechaczek C., Fetzer C., Baiker A., Bode J. and Lipps H.J. (1999). A vector based on the SV40 origin of replication and chromosomal S/MARs replicates episomally in CHO cells. *Nucl. Acids Res.* **27**, 426-428.

Portlock J.L. and Calos M.P. (2003). Site-specific genomic strategies for gene therapy. *Curr. Opin. Mol. Ther.* **5**, 376-382.

Pouton C.W. and Seymour L.W. (2001). Key issues in non-viral gene delivery. *Adv. Drug Delivery Rev.* **46**, 187-203.

Pruss D., Reeves R., Bushman F.D. and Wolffe A.P. (1994). The influence of DNA and nucleosome structure on integration events directed by HIV integrase. *J. Biol. Chem.* **269**, 25031-25041.

References

Ragot T., Vincent N., Chafrey P., Vigne E., Gilgenkrantz H., Couton D., Cartaud J., Briand P., Kaplan J.C., Perricaudet M. and Kahn A. (1993). Efficient adenovirus-mediated transfer of a human minidystrophin gene to skeletal muscle of mdx mice. *Nature* **361**, 647-650.

Raper S.E., Chirmule N., Lee F.S., Wivel N.A., Bagg A., Gao G.P., Wilson J.M. and Batshaw M.L. (2003). Fatal systemic inflammatory response syndrome in an ornithine transcarbamylase deficient patient following adenoviral gene transfer. *Mol. Genet. Metab.* **80**, 148-158.

Rausch H. and Lehmann M. (1991). Structural analysis of the actinophage PhiC31 attachment site. *Nucl. Acids Res.* **19**, 5187-5189.

Raymond C.S. and Soriano P. (2007). High-efficiency FLP and PhiC31 site-specific recombination in mammalian cells. *PLOS ONE* **2**, 1-4.

Reed R.R., Shibuya G.I. and Steitz, J.A. (1982). Nucleotide sequence of γδ resolvase gene and demonstration that its gene product acts as a repressor of transcription. *Nature* **300**, 381-383.

Rice P.A. (2005). Resolving integral questions in site-specific recombination. *Nat. Struct. Mol. Biol.* **12**, 641-643.

Rice P.A. and Steitz T.A. (1994). Model for a DNA-mediated synaptic complex suggested by crystal packing of γδ resolvase subunits. *EMBO J.* **13**, 1514-1524.

Rolland A. (2006). Nuclear gene delivery: the Trojan horse approach. *Exp. Opinion on Drug Delivery* **3**, 1-10.

Rowley P.A. and Smith M.C. (2008). Role of the N-terminal domain of PhiC31 integrase in attB-attP synapsis. *J. Bacteriol.* **190**, 6918-6921.

Rowley P.A., Smith M.C., Younger E. and Smith M.C. (2008). A motif in the C-terminal domain of PhiC31 integrase controls the directionality of recombination. *Nucl. Acids Res.* **36**, 3879-3891.

Saiki R.K., Gelfand D.H., Stoffel S., Scharf S.J., Higuchi R., Horn G.T., Mullis K.B. and Erlich H.A. (1988). Primer-directed enzymatic amplification of DNA. *Science* **239**, 487-491.

Sarkis G.J., Murley L.L., Leschziner A.E., Boocock M.R., Stark W.M. and Grindley N.D. (2001). A model for the gamma delta resolvase synaptic complex. *Mol. Cell* **8**, 623-631.

Sauer B. (1994). Site-specific recombination: developments and applications. *Curr. Opin. Biotechnol.* **5**, 521-527.

Schwikardi M. and Dröge P. (2000). Site-specificc recombination in mammalian cells catalyzed by γδ resolvase mutants: implications for the topology of episomal DNA (2000). *FEBS Letters* **471**, 147-150.

References

Schroeder A.R., Shinn P., Chen H., Berry C., Ecker J.R. and Bushman F. D. (2002). HIV-1 integration in the human genome favors active genes and local hotspots. *Cell*, **110**, 521-529.

Sclimenti C.R., Thyagarajan B. and Calos M.P. (2001). Directed evolution of a recombinase for improved genomic integration at a native human sequence. *Nucl. Acids Res.* **29**, 5044-5051.

Sharma N., Moldt B., Dalsgaard T., Jensen T.G. and Mikkelsen J.G. (2008). Regulated gene insertion by steroid-induced PhiC31 integrase. *Nucl. Acids Res.* **36**, e67. 1-12.

Simon R.H., Engelhardt J.F., Yang Y., Zepeda M., Weber-Pendleton S., Grossman M. and Wilson J.M. (1993). Adenovirus-mediated transfer of the CFTR gene to lung of nonhuman primates: toxicity study. *Hum Gene Ther.* **4**, 771-780.

Smith M.C., Till R., Brady K., Soultanas P., Thorpe H. and Smith M.C. (2004). Synapsis and DNA cleavage in phiC31 integrase-mediated site-specific recombination. *Nucl. Acids Res.* **32**, 2607-2617.

Smith M.C. and Thorpe H.M. (2002). Diversity in the serine recombinases. *Mol. Microbiol.* **44**, 299-307.

Sorrell D.A. and Kolb A.F. (2005). Targeted modification of mammalian genomes. *Biotechnol. Adv.* **23**, 431-469.

Stark W.M., Boocock M.R. and Sherratt D.J. (1992). Catalysis by site-specific recombinases. *Trends Genet.* **8**, 432-439.

Stark W.M., Sherratt D.J. and Boocock M.R. (1989). Site-specific recombination by Tn3 resolvase: topological changes in the forward and reverse reactions. *Cell* **58**, 779-790.

Sternberg N. (1979). A characterization of bacteriophage P1 DNA fragments cloned in lambda vector. *Virology* **96**, 129-142.

Stoll S.M., Ginsburg D.S. and Calos M.P. (2002). Phage TP901-1 site-specific integrase functions in human cells. *J. Bacteriol.* **184**, 3657-3663.

Struhl G. and Basler K. (1993). Organizing activity of wingless protein in Drosophila. *Cell* **72**, 527-540.

Thomas C.E., Ehrhardt A. and Kay M.A. (2003). Progress and problems with the use of viral vectors for gene therapy. *Nat. Rev. Genet.* **4**, 346-58.

Thomas C.E., Schiedner G., Kochanek S., Castro M.G. and Lowenstein P.R. (2001). Preexisting antiadenoviral immunity is not a barrier to efficient and stable transduction of the brain, mediated by novel high-capacity adenovirus vectors. *Hum. Gene Ther.* **12**, 839-846.

Thorpe H.M., Wilson S.E. and Smith M.C. (2000). Control of directionality in the site-specific recombination system of the streptomyces phage PhiC31. *Mol. Microbiol.* **38**, 232-241.

References

Thorpe H.M. and Smith M.C. (1998). In vitro site-specific integration of bacteriophage DNA catalyzed by a recombinase of the resolvase/ invertase family. *Proc. Nat. Acad. Sci. USA* **95**, 5505-5510.

Thyagarajan B., Liu Y., Shin S., Lakshmipathy U., Scheyhing K., Xue H., Ellerström C., Strehl R., Hyllner J., Rao M.S. and Chesnut J.D. (2008). Creation of engineered human embryonic stem cell lines using phiC31 integrase. *Stem Cells* **26**, 119-126.

Thyagarajan B. and Calos M.P. (2005). Site-specific integration for highlevel protein production in mammalian cells. In "Therapeutic Proteins: Methods and Protocols" (C. M. Smales, and D. C. James, Eds.), *Humana Press, Totowa, NJ*. 99-106.

Thyagarajan B., Olivares E.C., Hollis R.P., Ginsburg D.S. and Calos M.P. (2001). Site-specific genomic integration in mammalian cells mediated by phage PhiC31 integrase. *Mol. Cell. Biol.* **21**, 3926-3934.

Thyagarajan B., Guimaraes M.L., Groth A.C. and Calos M.P. (2000). Mammalian genomes contain active recombinase recognition sites. *Gene* **22**, 47-54.

Vega M.A. (1991). Prospects for homologous recombination in human gene therapy. *Hum. Gent.* **87**, 245-253.

Vigdal T.J., Kaufman C.D., Izsvák Z., Voytas D.F. and Ivics Z. (2002). Common physical properties of DNA affecting target site selection of Sleeping Beauty and other Tc1/ mariner transposable elements. *J. Mol. Biol.* **323**, 441-452.

Wang B.Y., Xu G.L., Zhou C.H., Tian L., Xue J.L., Chen J.Z. and Jia W. (2009). PhiC31 integrase interacts with TTRAP and inhibits NFkappaB activation. *Mol. Biol. Rep.* [Epub ahead of print]

Warren D., Lee S.Y. and Landy A. (2005). Mutations in the amino-terminal domain of lambda-integrase have differential effects on integrative and excisive recombination. *Mol. Microbiol.* **55**, 1104-1112.

Williams D.A. (2008). Sleeping beauty vector system moves toward human trials in the United States. *Mol. Ther.* **16**, 1515-1516.

Wilson M.H., Kaminski J.M. and George A. L. (2005). Functional zinc finger/ sleeping beauty transposase chimeras exhibit attenuated overproduction inhibition. *FEBS Letters* **579**, 6205-6209.

Wolff J.A., Malone R.W., Williams P., Chong W., Acsadi G., Jani A. and Felgner P.L. (1990). Direct gene transfer into mouse muscle *in vivo*. *Science* **247**, 1465-1468.

Woodard L.E., Hillman R.T., Keravala A., Lee S. and Calos M.P. (2009). Effect of nuclear localization and hydrodynamic delivery-induced cell division on PhiC31 integrase activity. *Gene Ther.* Oct 22. [Epub ahead of print]

Xu T. and Rubin G.M. (1993). Analysis of genetic mosaics in developing and adult Drosophila tissues. *Development* **117**, 1223-1237.

References

Yagil E., Dolev S., Oberto J., Kislev N., Ramaiah N. and Weisberg R.A. (1989). Determinants of site-specific recombination in the lambdoid coliphage HK022. An evolutionary change in specificity. *J. Mol. Biol.* **207**, 695-717.

Yang H.Y., Kim Y.W. and Chang H.I. (2002). Construction of an integration-proficient vector based on the site-specific recombination mechanism of enterococcal temperate phage PhiFC1. *J. Bacteriol.* **184**, 1859-1864.

Yang W. and Steitz T.A. (1995). Crystal structure of the site-specific recombinase γδ resolvase complexed with a 34 bp cleavage site. *Cell* **82**, 193-207.

Yang N.S. and Sun W.H. (1995). Gene gun and other non-viral approaches for cancer gene therapy. *Nat. Med.* **1**, 481-483.

Yant S.R., Huang Y., Akache B., and Kay M.A. (2007). Site-directed transposon integration in human cells. *Nucl. Acids Res.* **35**(7): e50.

Yant S.R., Wu X., Huang Y., Garrison B.S., Burgess S.M. and Kay M.A. (2005). High-resolution genome-wide mapping of transposon integration in mammals. *Mol. Cell. Biol.* **25**, 2085-2094.

Yant S.R., Park J., Huang Y., Mikkelsen J.G. and Kay M.A. (2004). Mutational analysis of the N-terminal DNA-binding domain of sleeping beauty transposase: critical residues for DNA binding and hyperactivity in mammalian cells. *Mol. Cell. Biol.* **24**, 9239-9247.

Yant S.R., Meuse L., Chiu W., Ivics Z., Izsvák Z. and Kay M.A. (2000). Somatic integration and long-term transgene expression in normal and haemophilic mice using a DNA transposon system. *Nat. Genet.* **25**, 35-41.

Yates J.L. and Guan N. (1991). Epstein-Barr virus-derived plasmids replicate only once per cell cycle and are not amplified after entry into cells. *J. Virol.* **65**, 483-488.

Yuan P., Gupta K. and Van Duyne G.D. (2008). Tetrameric structure of a serine integrase catalytic domain. *Strucutre* **6**, 1275-1286.

Zayed H., Izsvák Z., Walisko O. and Ivics Z. (2004). Development of hyperactive Sleeping Beauty transposon vectors by mutational analysis. *Mol. Ther.* **9**, 292-304.

Zhang M.X., Li Z.H., Fang Y.X., Zhu H.Z., Xue J.L., Chen J.Z. and Jia W. (2009). TAT-phiC31 integrase mediates DNA recombination in mammalian cells. *J. Biotechnol.* **142**, 107-113.

Zhang G., Gao X., Song Y.K., Vollmer R., Stolz D.B., Gasiorowski J.Z., Dean D.A. and Liu D. (2004). Hydroporation as the mechanism of hydrodynamic delivery. *Gene Ther.* **11**, 675-682.

Zhang G., Budker V. and Wolff J.A. (1999). High levels of foreign gene expression in hepatocytes after tail vein injections of naked plasmid DNA. *Hum. Gene Ther.* **10**, 1735-1737.

8. Appendix

Selection of amino acids discussed within this study.

Table 8.1. Chemical and physical properties and further characteristics of selected amino acids are presented, which are involved in the mutagenesis approach.

Amino acid	Chemical and physical properties	Further characteristics
Alanine; Ala (A)	aliphatic, non-polar,	tiny, ambivalent, can be inside or outside of the protein molecule
Aspartic acid Asp (D)	acidic, polar (negatively charged)	play important roles as general acids in enzyme active centers, as well as in maintaining the solubility and ionic character of proteins
Glutamic acid Glu (E)	acidic, polar (negatively charged)	
Lysine Lys (K)	basic, polar (positively charged)	frequently involved in salt bridges
Arginine Arg (R)	basic, polar (positively charged)	

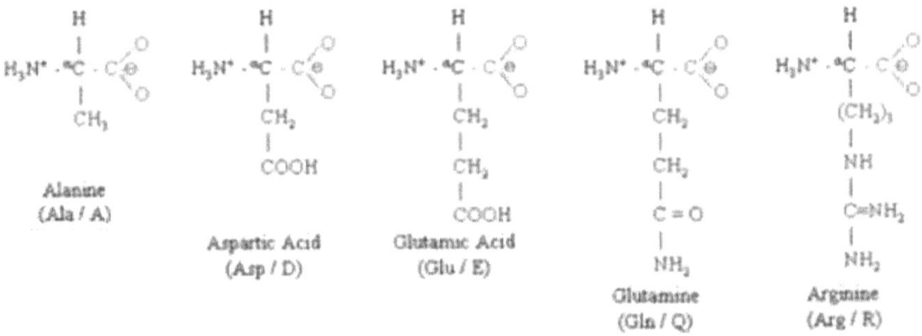

Figure 8.1. Selection of amino acids discussed within this study.
The four amino acids are shown, which were shifted to alanine in the selected mutagenesis approach alanine scanning. The amino group (-NH_3^+) is presented on the left of the amino acid and the carboxy group (COO^-) is depicted on the right site, respectively. The side chain for each amino acid is presented below the chiral C.atom in the middle.

Amino acid properties and structures were derived from the websites below:
1.) http://www.biology.arizona.edu/biochemistry/problem_sets/aa/aa.html
2.) http://www.thescienceforum.com/viewtopic.php?p=204149

Publications

Articles

Liesner R. et al.

Critical amino acid residues within the PhiC31 integrase DNA binding domain affect recombination activities in mammalian cells.

Human Gene Therapy, (2010) Sep. 21(9):1104-18.

Presentations of posters and talks with abstracts

05/2010	13th Annual meeting of the American Society of Gene and Cell Therapy Washington DC, USA
05/2008	11th Annual meeting of the American Society of Gene and Cell Therapy Boston, USA
12/2007	1st PhD symposium <interact> Life Science Community Munich Martinsried, Germany
07/2007	14th Annual meeting of the Deutsche Gesellschaft für Gentheraphie Heidelberg, Germany

Acknowledgements

First of all, I want to thank Prof. Dr. Heinrich Leonhardt for taking over supervision and representing this thesis at the Faculty of Biology at Ludwig-Maximilians-Universität München. I also want to thank PD Dr. Berit Jungnickel for writing the second opinion. Thanks as well to Prof. Dr. Schüler and Prof. Dr. Parniske and Prof. Dr.Eick for being a fair committee. Thanks also to my supervisor for giving me the opportunity to work in the laboratory.

I would also like to thank the entire working group at Max von Pettenkofer Institute I was working with, currently consisting of Nadine, Christina, Wenli, Martin, Nadja and Richard, who accounted for a lively and friendly atmosphere in the laboratory. Thank you for the nice time together in and out of the lab. I also greatly appreciate the help of Rudi for explaining me how to use the FACS machine. Additionally, thanks to Hans for scientific and mental support.

Thanks as well to Lorenz for relaxing lunch breaks and for his quiet manner, for his help and scientific expertise, I could benefit from. I also want to mention Nici. Thanks for late evening company after a laborious and long day in the lab. Thanks for lively conversations and support! Thanks for delicious home made exotic food supplying the whole team at lunch times.

I also want to thank my close friend Vijay for good conversations and support until the end and Klaus B., both also scientists and as well Robert for his mentoring and Mrs. Emmert for her large mental support beyond science.

Finally, I want to thank my family, my parents, my sister and my brother who never stopped their support and devoted much time to let this story and PhD thesis come to a good end. I specially want to thank my own family, my beloved partner Sabine and cute little Sarah (time bandit and motivator) for their endless support, motivation and critics and their love in all extents.

I want morebooks!

Buy your books fast and straightforward online - at one of world's fastest growing online book stores! Environmentally sound due to Print-on-Demand technologies.

Buy your books online at
www.morebooks.shop

Kaufen Sie Ihre Bücher schnell und unkompliziert online – auf einer der am schnellsten wachsenden Buchhandelsplattformen weltweit! Dank Print-On-Demand umwelt- und ressourcenschonend produziert.

Bücher schneller online kaufen
www.morebooks.shop

KS OmniScriptum Publishing
Brivibas gatve 197
LV-1039 Riga, Latvia
Telefax +371 686 204 55

info@omniscriptum.com
www.omniscriptum.com

Printed by Books on Demand GmbH, Norderstedt / Germany